Beyond the Spray: Delving into Diesel Engine Wall Interaction

Nicholas

First Printing, 2024

Contents

Chapter 1

Introduction

In this chapter, the motivation for understanding and studying the spray-wall interaction phenomenon in diesel engines, is introduced. First, a background in this field is presented in order to put this book in context. Then, the objectives that it pursues and the followed methodologies to achieve these purposes are described. Finally, the chapter outline that this document follows is portrayed.

1.1 Background and general context

The concern of the international community for environmental protection has increased, driven by the abrupt rise in energy consumption which is tied to the technological and social advancements over last decades. A field of special interest in this matter is covered by diesel engines, which find wide spread use for transportation, construction equipment, and countless industrial applications. This issue has been managed by governments through ever more stringent emissions regulations (such as the European Euro VI [1] or the Tier 4 U.S. standard [2] in the case of non-road engines), which forces the different manufacturers and researchers to find technical solutions to design efficient and cleaner engines that can still be economically feasible.

In this relentless search for technological innovations for diesel internal combustion engines (diesel ICEs), both computational and experimental progresses have been achieved, reaching a substantially great understanding of the phenomena involved in injection, spray development and combustion processes [3–7]. However, despite pushing forward ever more realistic and detailed modeling, the complete comprehension of the interaction between the spray and the piston walls is still elusive. The influence of the spray-wall interaction (SWI) on combustion is very complex. After the impingement, a film of fuel can be formed onto the walls, rising unburned hydrocarbons (UHC) and soot formation. Additionally, engine efficiency is reduced due, in a big extent, to the heat transfer from the reactive spray into the walls. The effect of the SWI on ignition occurrence timing in diesel engines is not clear either. On one hand, secondary atomization produced by the spray-wall impact enhances the droplet break-up, broadly reducing the drop size and fastening evaporation and combustion processes. On the contrary, the spreading of the spray on the wall would reduce local temperatures in the spray and would delay ignition. Therefore, the analysis of the spray-wall interaction is a key to get a better understanding of in-chamber processes in ICEs, specially when it occurs irremediably in small-bore diesel and direct injection gasoline engines and it influences even more at cold-start conditions, where low densities and temperatures promote liquid penetration of the spray until it reaches the walls.

Several authors [8, 9] have classified the analysis of SWI according to the morphologhy of the colliding fuel: The study of single drops or impinging sprays onto walls. Even though the isolated droplet impact is simplified respect to the impinging spray one, the single-droplet-based literature has made a substantial contribution to the understanding of the influence of a big amount of variables on spray-wall interaction, such as the properties of the fluid and the surface and the characteristics of formation and impingement of the droplets [10, 11]. Computational fluid dynamics (CFD) simulations are an appealing way to study SWI, due to the difficulty of assessing to the engine through experimental techniques, and the cost reduction that their use implies. One of the first attempts to model spray-wall interaction, which in that time presented very good agreement with the spray contours provided by Kuniyoshi et al. in [12], was introduced by Naber and Reitz in 1988 [13], who simulated the drops interaction with walls using three different models of behavior: Stick or adhesion, Reflect or specular rebound and Jet or tangential incidence. This study has been a relevant starting point to other modern and complex computational works [11, 14–22].

Nevertheless, the accuracy of numerical models is generally supported on the presence of reliable high-quality data obtained experimentally. Again, the two different approaches between isolated drops [10, 11, 23, 24] and sprays in a scant extent [17, 25–27] has been explored in experimental investigation. Fast temperature probes have been used by several researchers [25, 28, 29] to study heat transfer processes between the diesel spray and solid surfaces, getting interesting results such as the effect of pre-impingement spray velocities on the local heat transfer rates. Variables to be taken into account are the ambient conditions where the fuel is injected. Akop et al. studied the spray characteristics in an ambient at atmospheric pressure [30], while a constant-volume vessel was set at 4.2 MPa and 1000 K by López and Pickett in [31] with the purpose of analyze the effect of SWI on soot formation. Those aforementioned experiments were carried out injecting fuel on a flat wall. Another approximation employed by Pickett and López in a different research [32] was to confine the jet inside a flat wall and transparent side walls in order to simulate a piston bowl for a typical heavy-duty combustion chamber near top dead center. As seen, another kind of classification in the studies on this matter is the similarity of the environment with a combustion chamber, both in terms of the wall geometry and the gas thermodynamic conditions. However, the more the experiment resembles real diesel in-cylinder conditions, the more difficult is to assess the individual effect of a determined parameter on SWI. For that reason, less realistic situations are often used [24, 30, 33].

This investigation contributes to the current understanding of the spray-wall interaction using an intermediate approach, in which real injectors are tested at a wide span of realistic in-cylinder conditions of high-pressure and high-temperature. Nonetheless, the wall has been simplified to be a flat plate in order to insulate the studied phenomenon and extract as much information as possible about the influence of the ambient, injection and wall-positioning conditions on the SWI phenomenon, in terms of air-fuel mixture, spreading along the surface, combustion parameters and spray-wall heat exchange.

1.2 Objectives and methodology

This book aims to elucidate the aforementioned spray-wall interaction issues and determine the key parameters who control this phenomenon and their influence on atomization, evaporation and combustion processes. This work was made mainly in the frame of two research collaborative projects between the Injection and Combustion Research Group of CMT-Motores Térmicos and Caterpillar Inc., specifically the Innovation and Technology Development Division. Both projects are originally based on experimental data acquisition of the injection and spray impact with a flat wall at different positions and diesel-like conditions, essentially with CFD models validation and fundamental understanding purposes. The main differences between both projects are based in the employed wall hardware and the utilization of fast response thermocouples to include heat transfer to the wall in the analysis. With the objective of widening the range of study, one additional fuel and one injector, along with more operating and wall positioning conditions, were included in the test plans by CMT-Motores Térmicos.

The Engine Combustion Network (ECN) is a worldwide forum in which several research institutions focus efforts on the experimental and computational analysis of determined cases of study within the engine combustion field [34–38]. One of the main goals of the ECN is the generation of a high-quality and reliable open database (available in `https://ecn.sandia.gov/` [39]). Single and multi-orifice injectors (donated by Robert Bosch GmbH and Delphi Automotive Systems) are target hardware for the ECN activities. Therefore, due to its relevance for the scientific community, a solenoid-driven single-hole injector from ECN referred to as 'Spray D' was taken as reference in this book. The manufacture of its nozzle makes it prone to suppress internal flow cavitation [40].

Having said all that, the purposes of the present book could be summed up in the following points:

- To contribute in the current understanding and determination of the influence of the induced variations in ambient, injection and wall conditions on the macroscopic evolution of the spray, combustion characteristics and heat transfer when it collides with a wall.
- The acquisition of experimental data of spray-wall interaction at realistic engine conditions through advanced diagnosis techniques and modern equipment, in order to extract valuable information for CFD modeling development and validation. In this regard, the simplification of the experiment using a flat wall is pivotal in order to assess the different parameters that rule SWI with reasonable complexity and similarity to in-chamber conditions (evaporative and reacting sprays).

The experimental work was carried out entirely taking advantage of the state-of-the-art hardware available at CMT-Motores Térmicos. Particularly, a high-pressure and high-temperature vessel has been used to reproduce the ambient conditions present in a diesel engine. Two different fuels (n-dodecane and commercial diesel) have been injected within this facility. The evaporative inert spray was recorded at high speed, getting simultaneously vapor and liquid phases images. Similarly, the spray at reactive conditions was recorded to get the flame evolution from different angles, along with the vapor phase evolution during the injection event and the lift-off length through OH* chemiluminescence. In regards to the wall, two different systems were developed and used: a transparent quartz wall with nearly initial isothermal conditions ($T_w \approx T_{amb}$) and an instrumented and cooled stainless steel wall provided with novel fast-response thermocouples.

Aside the equipment and facilities, the dense experience of the *Departamento de Máquinas y Motores Térmicos* at the *Universitat Politècnica de València* (to which CMT is attached) in the study of diesel injection systems has defined important precedents to this work. Numerous PhD theses presented in the last years have not only been useful to define relevant inputs for CFD model validation [41–45], but also to introduce some theoretical argumentations and the use of characterization experimental techniques, important to the understanding of the hydraulic behavior of the injector used in this work [46–49]. Some other theses even presented the operation of the same test rig and similar physical principles than the used in this document [50–53].

Nevertheless, these previous researches, with the exception or the González's PhD book [46] which is the one closely related to spray-wall interaction, are focused in the study of sprays at free-jet conditions. González took into account several possibilities in regards to the impingement angle between the spray and the wall. He established a model to predict the spray spreading evolution along the wall and also a law of distribution of drop sizes through the analysis macroscopic and microscopic of SWI. However, this work did not considered the evaporation and combustion processes, limiting itself to the study of non-evaporative jet in a chamber at ambient temperature. This simplification was suppressed by the use of the test vessel mentioned above employed in four previous doctoral theses [50–53]. However, the facility had to be adapted for the first time to the study of SWI through the design and implementation of suitable wall systems, which will be discussed in this manuscript.

1.3 About this book

The document is divided into 8 chapters, starting with the present introduction (**chapter 1**), that serves to briefly describe the general context and the main objectives of the book.

In **chapter 2**, the fundamental concepts of diesel direct injection are discussed. Different technologies on this topic are presented along with the different processes that take place into the combustion chamber during the fuel injection and a short introduction of the macroscopic spray development is given. Finally, spray-wall interaction is presented in the context of internal combustion engines and a review of different findings found in the literature is presented.

Next, **chapter 3** follows with the description of the facilities, equipment and general tools employed in this work, emphasizing on the injection system, the high-temperature and high-pressure vessel and the two different wall systems that were used to simulate spray-wall interaction at diesel-like conditions. Relevant properties of the fuels are also provided.

The experimental and processing methodologies are outlined in **chapter 4**, providing a detailed description of the techniques principles, the setups employed during each campaign and the data treatment of both images and voltage signals.

The following three chapters are focused on the presentation and discussion of the experimental results. Through **chapter 5**, results of the evaporative inert spray visualization are shown separated into vapor and liquid phases and are subdivided into sections before and after-impingement regions, to finally proposes empirical models from those results. Reacting spray interaction with the transparent quartz wall is encompassed in **chapter 6**, in terms of different parameters such as ignition delay, spray penetration, spray spreading and thickness along the wall and lift-off length, discussing the observed trends. In **chapter 7** ignition conditions are presented, now with the tempered wall and the effects of plate temperature and heat flux, and its estimation is added to the analysis of the spray-wall interaction characteristics.

Finally, **chapter 8** closes the book outline by stating the main conclusions extracted from the results and discussions previously made and, additionally, presents future work proposals to go further in the study of the topic covered by this manuscript.

References

[1] *On type-approval of motor vehicles and engines with respect to emissions from heavy duty vehicles (Euro VI) and on access to vehicle repair and maintenance information and amending Regulation (EC) No 715/2007 and Directive 2007/46/EC and repealing Directi.* Official Journal of the European Union. Regulation. 2009 (cited on page 1).

[2] *Control of Emissions of Air Pollution From Nonroad Diesel Engines and Fuel; Final Rule.* Environmental Protection Agency. Regulation. 2004 (cited on page 1).

[3] Higgins, Brian and Siebers, Dennis L. "Measurement of the Flame Lift-Off Location on DI Diesel Sprays Using OH Chemiluminescence". In: *SAE Paper 2001-01-0918* (2001) (cited on page 2).

[4] Kastengren, Alan L et al. "Engine Combustion Network (ECN): Measurements of Nozzle Geometry and Hydraulic Behavior". In: *Atomization and Sprays* 22.12 (2012), pp. 1011–1052 (cited on page 2).

[5] Xue, Qingluan et al. "Eulerian CFD Modeling of Coupled Nozzle Flow and Spray with Validation Against X-Ray Radiography Data". In: *SAE International Journal of Engines* 7.2 (2014), pp. 1061–1072 (cited on page 2).

[6] Montanaro, Alessandro et al. "Schlieren and Mie Scattering Visualization for Single- Hole Diesel Injector under Vaporizing Conditions with Numerical Validation". In: *SAE Technical Paper* (2014) (cited on page 2).

[7] Payri, Raul, Salvador, Francisco Javier, Gimeno, Jaime, and Peraza, Jesús E. "Experimental study of the injection conditions influence over n-dodecane and diesel sprays with two ECN single-hole nozzles. Part II: Reactive atmosphere". In: *Energy Conversion and Management* 126 (2016), pp. 1157–1167 (cited on page 2).

[8] Tropea, Cameron and Marengo, Marco. "The Impact of Drops on Walls and Films". In: *Multiphase Science and Technology* 11 (1998), pp. 11–36 (cited on page 2).

[9] Lee, Sang Yong and Ryu, Sung Uk. "Recent progress of spray-wall interaction research". In: *Journal of Mechanical Science and Technology* 20.8 (2006), pp. 1101–1117 (cited on page 2).

[10] Panão, M.R.O. and Moreira, A.L.N. "Experimental study of the flow regimes resulting from the impact of an intermittent gasoline spray". In: *Experiments in Fluids* 37 (2004), pp. 834–855 (cited on page 2).

[11] Moreira, A.L.N., Moita, A S, and Panão, Miguel R. O. "Advances and challenges in explaining fuel spray impingement : How much of single droplet impact research is useful ?" In: *Progress in Energy and Combustion Science* 36 (2010), pp. 554–580 (cited on page 2).

[12] Kuniyoshi, H, Tanabe, H, Sato, G T, and Fujimoto, Hajime. "An Investigation on the characteristics of Diesel fuel spray". In: *SAE Paper 800968* (1980) (cited on page 2).

[13] Naber, J. and Reitz, R. "Modeling Engine Spray/Wall Impingement". In: *SAE Technical Paper 880107* (1988) (cited on page 2).

[14] Yarin, A. L. and Weiss, D. A. "Impact of drops on solid surfaces: self-similar capillary waves, and splashing as a new type of kinematic discontinuity". In: *Journal of Fluid Mechanics* 283 (1995), pp. 141–173 (cited on page 2).

[15] Cossali, G E, Coghe, A, and Marengo, M. "The impact of a single drop on a wetted solid surface". In: *Experiments in Fluids* 22 (1997), pp. 463–472 (cited on page 2).

[16] Habchi, C, Foucart, H, and Baritaud, T. "Influence of the Wall Temperature on the Mixture Preparation in DI Gasoline Engines". In: *Oil & Gas Science and Technology - Revue d'IFP* 54.2 (1999), pp. 211–222 (cited on page 2).

[17] Trujillo, M. F., Mathews, W. S., Lee, C. F., and Peters, J. E. "Modelling and experiment of impingement and atomization of a liquid spray on a wall". In: *International Journal of Engine Research* 1.1 (2000), pp. 87–105 (cited on page 2).

[18] Lippert, Andreas M, Stanton, Donald W, Reitz, Rolf D, Rutland, Christopher J, and Hallett, William L H. "Investigating the Effect of Spray Targeting and Impingement on Diesel Engine Cold Start". In: *SAE Technical Paper 2000-01-0269* (2000) (cited on page 2).

[19] Cole, Vernon, Mehra, Deepak, Lowry, Sam, and Gray, Donald. "A Numerical Spray Impingement Model Coupled With A Free Surface Film Model". In: *The Fifth International Symposium on Diagonstics and Modeling of Combustion in Internal Combustion Engines.* 2001 (cited on page 2).

[20] Montorsi, Luca, Magnusson, Alf, and Andersson, Sven. "A Numerical and Experimental Study of Diesel Fuel Sprays Impinging on a Temperature Controlled Wall". In: *SAE Technical Paper 2006-01-3333* 724 (2006), pp. 776–790 (cited on page 2).

[21] Jia, Ming, Peng, Zhijun, Xie, M, and Stobart, R. "Evaluation of spray/wall interaction models under the conditions related to diesel HCCI engines". In: *SAE Technical Papers* 1.1 (2008), pp. 993–1008 (cited on page 2).

[22] Yu, H. et al. "Numerical Investigation of the Effect of Alcohol-Diesel Blending Fuels on the Spray-Wall Impingement Process". In: *SAE Technical Papers* 2016-April.April (2016) (cited on page 2).

[23] Hardalupas, Y, Okamoto, S, Taylor, AMKP, and Whitelaw, JH. "Application of a phase Doppler anemometer to a spray impinging on a disc". In: *Laser techniques and applications in fluid mechanics, Proc 6th Int Symp. Berlin.* Ed. by Durao D Adrian R and Maeda M Durst F, Heitor M. Springer-Verlag, 1992, pp. 490–506 (cited on page 2).

[24] Ko, Kyungnam and Arai, Masataka. "Diesel Spray Impinging On A Flat Wall, Part I: Characteristics Of Adhered Fuel Film In An Impingement Diesel Spray". In: *Atomization and Sprays* 12.5&6 (2002), pp. 737–751 (cited on pages 2, 3).

[25] Arcoumanis, Constantine and Chang, J. C. "Heat transfer between a heated plate and an impinging transient diesel spray". In: *Experiments in Fluids* 16.2 (1993), pp. 105–119 (cited on page 2).

[26] Pan, H. et al. "Experimental Investigation of Fuel Film Characteristics of Ethanol Impinging Spray at Ultra-Low Temperature". In: *SAE Technical Paper 2017-01-0851* (2017) (cited on page 2).

[27] Zhao, L. et al. "An Experimental and Numerical Study of Diesel Spray Impingement on a Flat Plate". In: *SAE Int. J. Fuels Lubr.* 10.2 (2017), pp. 407–422 (cited on page 2).

[28] Arcoumanis, C, Cutter, P, and Whitelaw, D S. "Heat Transfer Processes In Diesel Engines". In: *Trans IChemE* 76.2 (1998), pp. 124–132 (cited on page 2).

[29] Mahmud, Rizal et al. "Characteristics of Flat-Wall Impinging Spray Flame and Its Heat Transfer under Small Diesel Engine-Like Condition". In: *SAE Technical Paper 2017-32-0032* (2017) (cited on page 2).

[30] Akop, Mohd Zaid, Zama, Yoshio, Furuhata, Tomohiko, and Arai, Masataka. "Characteristics Of Adhesion Diesel Fuel On An Impingement Disk Wall Part 1: Effect Of Impingement Area And Inclination Angle Of Disk". In: *Atomization and Sprays* 23.8 (2013), pp. 725–724 (cited on page 3).

[31] Lopez, Jose Javier and Pickett, Lyle M. "Jet/wall interaction effects on soot formation in a diesel fuel jet". In: *International Symposium on Diagnostics and Modeling of Combustion in Internal Combustion Engines (COMODIA)* (2004), pp. 387–394 (cited on page 3).

[32] Pickett, Lyle M and Lopez, Jose Javier. "Jet-wall interaction effects on diesel combustion and soot formation". In: 2005.724 (2005) (cited on page 3).

[33] Payri, Raul, Gimeno, Jaime, Peraza, Jesús E., and Bazyn, Tim. "Spray / wall interaction analysis on an ECN single-hole injector at diesel-like conditions through Schlieren visualization". In: *Proc. 28th ILASS-Europe, Valencia* September (2017) (cited on page 3).

[34] Pickett, Lyle M, Genzale, Caroline L, Manin, Julien, Malbec, Louis-Marie, and Hermant, Laurent. "Measurement Uncertainty of Liquid Penetration in Evaporating Diesel Sprays". In: *ILASS Americas 23rd Annual Conference on Liquid Atomization and Spray Systems*. Ventura, CA (USA): ILASS-Americas, 2011 (cited on page 4).

[35] Saha, Kaushik et al. "Numerical Investigation of Two-Phase Flow Evolution of In- and Near-Nozzle Regions of a Gasoline Direct Injection Engine During Needle Transients". In: *SAE International Journal of Engines* 9.2 (2016), pp. 2016–01–0870 (cited on page 4).

[36] Desantes, Jose Maria, Garcia-Oliver, Jose Maria, Pastor, Jose Manuel, and Pandal, Adrian. "A Comparison of Diesel Sprays CFD Modeling Approaches: DDM versus Σ-Y Eulerian Atomization Model". In: *Atomization and Sprays* 26.7 (2016), pp. 713–737 (cited on page 4).

[37] Skeen, Scott A et al. "A Progress Review on Soot Experiments and Modeling in the Engine Combustion Network (ECN)". In: *SAE International Journal of Engines* 9.2 (2016) (cited on page 4).

[38] Gimeno, Jaime, Bracho, Gabriela, Martí-Aldaraví, Pedro, and Peraza, Jesús E. "Experimental study of the injection conditions influence over n-dodecane and diesel sprays with two ECN single-hole nozzles. Part I: Inert atmosphere". In: *Energy Conversion and Management* 126 (2016), pp. 1146–1156 (cited on page 4).

[39] ECN. *Engine Combustion Network*. Online. 2010 (cited on page 4).

[40] Payri, Raul, Gimeno, Jaime, Cuisano, Julio, and Arco, Javier. "Hydraulic characterization of diesel engine single-hole injectors". In: *Fuel* 180 (2016), pp. 357–366 (cited on page 4).

[41] López, José Javier. "Estudio teórico-experimental del chorro libre Diesel no evaporativo y de su interacción con el movimiento del aire". PhD book. Valencia: E.T.S. Ingenieros Industriales. Universitat Politècnica de València, 2003 (cited on page 5).

[42] Salvador, Francisco Javier. "Influencia de la cavitación sobre el desarrollo del chorro Diesel". PhD book. E.T.S. Ingenieros Industriales. Universitat Politècnica de València, 2003 (cited on page 5).

[43] Martí-Aldaraví, Pedro. "Development of a computational model for a simultaneous simulation of internal flow and spray break-up of the Diesel injection process". PhD book. Valencia: Universitat Politècnica de València, 2014 (cited on page 5).

[44] Carreres, Marcos. "Thermal Effects Influence on the Diesel Injector Perfomance through a Combined 1D Modelling and Experimental Approach". PhD book. Universitat Politècnica de València, 2016 (cited on page 5).

[45] Jaramillo Císcar, David. "Estudio experimental y computacional del proceso de inyección diésel mediante un código CFD con malla adaptativa". PhD book. Universitat Politècnica de València, 2017 (cited on page 5).

[46] González, Uriel. "Efecto del choque de pared en las características del chorro Diesel de inyección directa". PhD book. Valencia: E.T.S. Ingenieros Industriales. Universitat Politècnica de València, 1998 (cited on page 5).

[47] García-Oliver, José María. "Aportaciones al estudio del proceso de combustión turbulenta de chorros en motores Diesel de inyección directa". PhD book. Valencia: E.T.S. Ingenieros Industriales. Universitat Politècnica de València, 2004 (cited on page 5).

[48] Gimeno, Jaime. "Desarrollo y aplicación de la medida de flujo de cantidad de movimiento de un chorro Diesel". PhD book. E.T.S. Ingenieros Industriales. Universitat Politècnica de València, 2008 (cited on page 5).

[49] Manin, Julien. "Analysis of mixing processes in liquid and vaporized diesel sprays through LIF and Rayleigh scattering measurements". PhD book. Valencia: E.T.S. Ingenieros Industriales. Universidad Politécnica de Valencia, 2011 (cited on page 5).

[50] Bardi, Michele. "Partial needle lift and injection rate shape effect on the formation and combustion of the Diesel spray". PhD book. Valencia (Spain): Universitat Politècnica de València, 2014 (cited on page 5).

[51] Viera, Juan Pablo. "Experimental Study of the Effect of Nozzle Geometry on the Performance of Direct-Injection Diesel Sprays for Three Different Fuels". PhD book. Universitat Politècnica de València, 2017 (cited on page 5).

[52] Vaquerizo, Daniel. "Study on Advanced Spray-Guided Gasoline Direct Injection Systems". PhD book. Universitat Politècnica de València, 2017 (cited on page 5).

[53] Giraldo Valderrama, Jhoan Sebastián. "Macroscopic and microscopic characterization of non-reacting diesel sprays at low and very high injection pressures". PhD book. Universitat Politècnica de València, 2018 (cited on page 5).

Chapter 2

The diesel injection and spray-wall interaction

The main goal of the present chapter is to provide an overview about the fundamentals on injection of diesel sprays and its interaction with walls, indicating the literature review which serves as a base for this work. The mechanisms employed in diesel engines for fuel injection and their functioning principles are briefly described. After that, the diesel spray is presented, covering the injection process and the main characteristics of the spray and how are they influenced by different parameters. Then, the diesel spray impact with walls is described through the progressive knowledge obtained in research about this phenomenon and its manifestation in internal combustion engines.

2.1 Diesel injection systems

The injection system is the responsible to provide fuel within the combustion chamber of the engine to its correct functioning. It has to work in a well-controlled way following the mass flow rate law imposed in order to supply the right amount of fuel when needed and manage the different stages of combustion. Additionally, the fuel has to be properly mixed with the in-chamber gases by incrementing their contact surface after the compression stroke. All these processes have to be in synchrony with the crankshaft-piston system movement. Taking into account its role in air-fuel mixture formation, its strong influence on engine performance, efficiency and pollutant formation can be inferred.

Depending on the way in which the fuel is delivered to the combustion chamber, diesel injection systems could be divided into two different categories: indirect injection and direct injection. Both hardware are drawn in Figure 2.1. In indirect injection engines, the combustion chamber has two main and differenced areas that are united by a small duct between them, the pre-combustion chamber and the main chamber. The fuel is injected through a single-hole nozzle into the pre-combustion chamber, once the fuel gets in contact with the surrounding air, an incomplete combustion is produced within this small chamber. The duct enhances air turbulence during the compression stroke, improving the mixing with fuel. This makes the function of fuel atomization more over the air than over the injector, which allows to use single-hole systems with big diameters (around 1 mm) and injection pressures below 50 MPa. On the contrary, direct injection (DI) systems just have one injection chamber which is machined in the piston and the fuel is directly injected there. The air is just induced by the intake conducts at relative low velocities, which compel to use more advanced technologies for injection systems than for indirect injection. In this regard, DI systems typically use multi-hole nozzles with injection pressures up to 300 MPa and with 100 µm diameter holes.

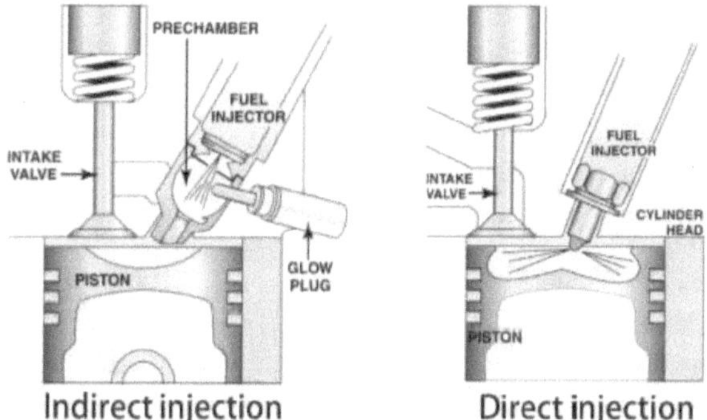

Fig. 2.1: Scheme of both indirect (left) and direct (right) diesel injection systems [1].

Due to a lower surface-volume ratio in direct injection engines and the lack of pressure losses in the indirect injection systems orifice, DI systems broadly have a higher efficiency. This fact has made that, currently, the indirect injection systems had been displaced as a design solution for commercial diesel engines by direct injection technologies.

2.1.1 Diesel direct injection systems

As mentioned, in direct injection systems the fuel is directly injected into the combustion chamber and it is the fuel spray itself the responsible for the mixture formation with the surroundings. Broadly-speaking, high-injection pressures and strong atomization regimes are the means for that. Below are listed some of the more employed systems and commercialized nowadays:

- **Rotary pump or direct actuation systems:** Those are the first systems used in direct injection diesel engines. They are conformed by a high-pressure pump that supplies the fuel to the injector, which has the only function of atomize it. The main drawback of this system is the impossibility of controlling both the injection pressure, which is dependent on the pump speed, and the timing for start of the injection, which happens when the injection pressure is high enough to displace the injector spring.

- **Injector-pump or unit injector:** This system puts together the fuel pump and the injector into one sole device, which is placed in the cylinder head and driven directly by the camshaft. A solenoid valve, activated by an Electronic Control Unit (ECU) is used to control the start and the duration of the injector, allowing additionally to disconnect different cylinders at partial load conditions. One particular case of this system is the one called 'unit pump', where the pump and the injector are connected by a short high-pressure line.

- **Accumulation systems:** Its functioning principle is based on the accumulation of fuel at a certain pressure in a deposit between the pump and the injector in order to absorb pressure oscillation produced by the pump and the fuel injection event. The pump is driven by the crankshaft with the sole purpose of rising the injection pressure of the system, while this set injection pressure and the opening of the injector are electronically controlled. The common-rail, that is described in more detail in the following subsection, is one of the accumulation systems whose implementation is more spread nowadays due to its advantages in terms of pollutant emissions and reduction en mechanical issues.

2.1.2 The common-rail system

As mentioned above, the common-rail is one of the most used accumulations systems, whose injector actuation is electronically controlled and allows to damper pressure oscillations. In Figure 2.2 a diagram of a typical common-rail arrangement can be found. A low-pressure pump aspire the fuel from the tank and deliver it to a high-pressure pump that is driven by the crankshaft. Then, the fuel is sent to the different injectors, being distributed from a common deposit (the rail or common-rail). The volume of the common-rail and the lines between the pump and the injector serves as accumulator, preventing fluctuations in the MPa order.

The fuel pressure in the rail is measured by a sensor and the signal is compared to a set-point value by the Electronic Control Unit, and any differences between both are regulated by a discharge orifice located in the high-pressure side of the system. The fuel that is extracted for this control, together with the fuel of the injectors return (used to control the opening of the needle and the return of the high-pressure pump) are connected and sent back to the fuel tank. The ECU also controls the opening and closing of the injectors, which together with the pressure governs the injection law and the amount of injected fuel.

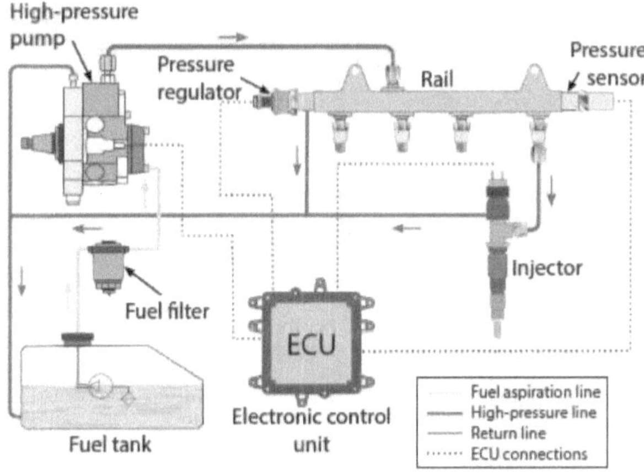

Fig. 2.2: Scheme of a generic common-rail system [2].

The common-rail injector is the most important and complex part of the whole injection system, which has made that most efforts have been focused on the improvement of this element. State-of-the-art injectors have raised the maximum injection pressure limits up to 300 MPa, in order to increase the capability of delivering fuel in a short period of time and enhance the air-fuel mixing. Depending on the actuator physical principle, common-rail injectors may be divided into different groups, where two of them are the most extensively spread technologies: solenoid driven or piezoelectric driven injectors.

2.1.3 Solenoid-driven injectors

Figure 2.3 presents the elements that compose a common-rail injector. They can be divided into three different sub-systems: servo-actuated system, which is composed by a control volume (CV) and a fuel valve; the connecting rod, which transmits the motion from the servo-actuated system to the injector needle and the injector nozzle, where the needle could also be included and it is responsible for controlling the fuel flow through the orifices. The servo-actuated system is the core of the solenoid-driven injector work principle.

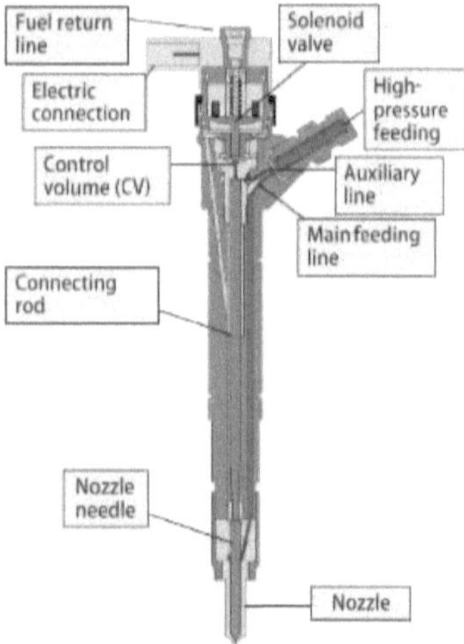

Fig. 2.3: Global sketch of a solenoid-driven common-rail injector (figure shows a Bosch CRI2.4) [3].

In Figure 2.4 can be observed how a solenoid-activated injector works: while in rest condition of the rod (top), the small ball valve is closed and there is not flow through the return. Fuel pressure in the control volume (CV) is the same as in the high-pressure line and the rail. As the area of the connecting rod is larger than the one of the needle, the balance of forces pushes it against the nozzle, keeping the injector closed. In the opening stage (Figure 2.4-center), the ECU delivers a signal whose intensity goes through the solenoid coils and generates a force that opens the ball valve. The duration of this pulse is known as energizing time (ET), and it is used to control the amount of injected fuel and the injection law. The flow through the volume control outlet orifice, and the losses in the inlet orifice (CO) produce a pressure drop in the CV, which unbalances the forces between the needle and the piston, causing a motion of the needle towards the control volume and, consequently opening the nozzle orifices and allowing the injection of fuel through them. From the beginning of the pulse to the moment when needle displacement allows the exit of fuel through the nozzle outlet, there is a delay which is known as start of injection or SOI. Finally, in the stage shown in the bottom part of Figure 2.4, the current is extinguished, producing the obstruction of the ball valve. The return flow is interrupted and the control volume is filled again with fuel from the rail (high-pressure). Again, the difference between the sections of the piston and the needle produces a force delta which sends the needle to the nozzle seat and closes the orifices, leading to the end of the injection.

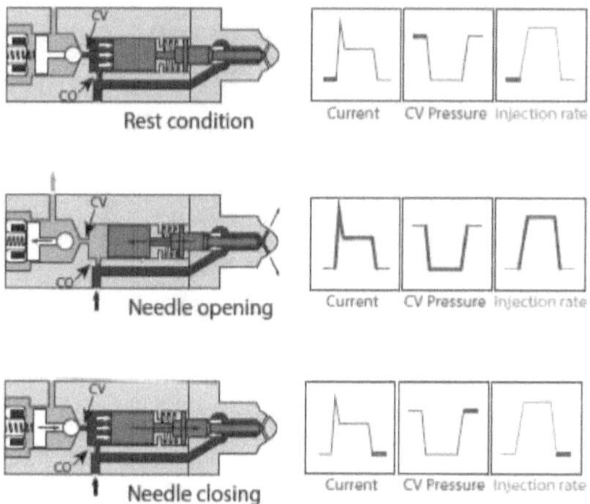

Fig. 2.4: Functioning principle of a solenoid-driven injector.

Having the control volume aligned with the needle improves the hydrodynamic response of the system and allows to do multiple injections with high accuracy. However, in order to enhance the response time of the injection, another technology which is widely used nowadays is piezoelectric-commanded injectors.

2.1.4 Piezoelectric-driven injectors

Even when the working principle of piezoelectric-driven injectors is quite similar to the above described, the difference lies on the actuator that, instead of the solenoid coil, is based on the stacking of quartz crystals between metallic sheets connected in series. When an electric voltage is applied to both sheets, the crystals are elongated and this variation of length can be used to activate a valve that reliefs or seals the control volume. The internal components of a piezoelectric-actuated injector are shown in Figure 2.5. It typically consists of the nozzle, the injector body and the piezoelectric valve, which is formed by a piezoelectric actuator, a hydraulic amplifier and a control valve, as detailed in Figure 2.5-right. In this case, two inlet orifices and one outlet orifice (OAZ), which also works as an inlet orifice.

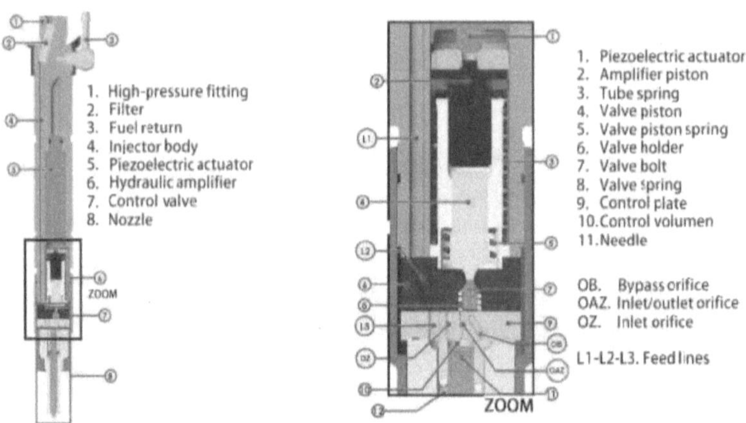

Fig. 2.5: Components of a piezoelectric-driven injector (figure shows a Bosch CRI3.1 injector). Left: transversal cut of the injector. Right: detailed zoom of the piezoelectric valve [4].

Due to the variation in length produced by a current in piezoelectric materials, the actuator expands when it is excited, displacing the amplifier piston and compressing the fuel between both extremes of the valve piston. This provokes a vertical displacement of the command piston, which opens the upper side of the control valve and closes the lower one, thus keeping the bypass orifice OB inactive. Once the upper side of the control valve is opened, the pressure downstream of OAZ is reduced, creating fuel flow through that orifice. The pressure drop in the control volume causes needle lift and the orifices are unblocked, resulting in fuel injection. When the actuator is deactivated, it returns to its original length by means of the tube spring. The upper side of the valve bolt closes and the pressure in the control volume is then risen with the fuel entrance through both OAZ and OZ orifices. This pressure recovery provokes the needle closing and the end of the injection.

Those systems are commonly faster than solenoid-driven injectors, being ideal for multiple injections (reaching about eight per cycle). Furthermore, they are highly accurate, controlling needle displacements in the order of nanometers by varying the voltage.

2.1.5 Diesel injector nozzles

The injector dynamics and the behavior of the spray in terms of external flow, air-fuel mixing and combustion, are strongly influenced by the flow conditions through the injector control volumes and the nozzle orifices [5–8]. For this reason, the injector nozzle characteristics are key parameter are key parameters that play a fundamental role in the diesel spray formation and all later processes. In Figure 2.6, two different types of nozzles, microsac (left) and valve closed orifice or VCO (right), are shown. These are the most widely used commercially. Microsac nozzles have a big advantage in regards to the control of the fuel flow without being strongly affected by the needle position during opening and closing transients, thanks to the sac volume. Nevertheless, this volume causes a problem of residual injection after needle closing or dribble, which could lead to soot and unburned hydrocarbons formation. On the other hand, VCO nozzles do not present this issue but the liquid flow is highly influenced by the needle movement.

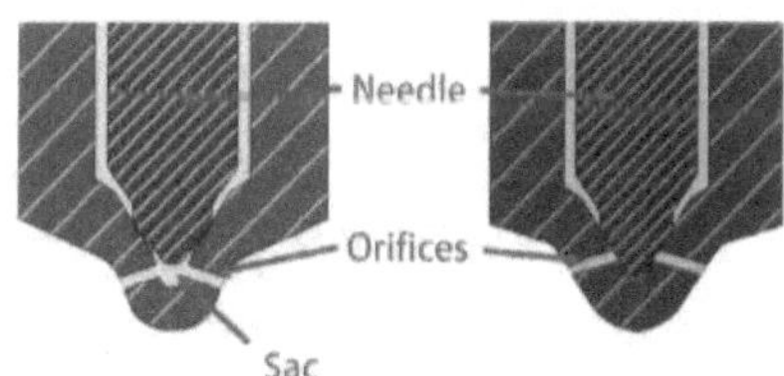

Fig. 2.6: Diesel nozzle types. Left: Microsac. Right: VCO.

When fuel is forced to go through the different injector channels and volumes, it finds discharge in the nozzle orifices, where the flow is accelerated accordingly to their geometric features, which has a great effect on the behavior of the spray. Figure 2.7 shows a scheme with the main characteristics of a nozzle: hole length (L_h), inlet diameter (D_i), outlet diameter (D_o) and inlet curvature radius (r_c).

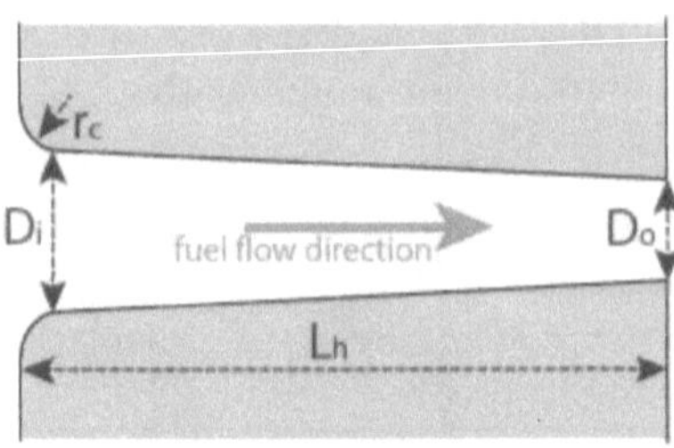

Fig. 2.7: Geometric parameters of a nozzle orifice.

Orifice convergence and inlet curvature can not only influence internal flow acceleration but can also produce pressure drops that can cause local phase changes, such as cavitation [8–10]. One parameter frequently used in literature and industry to characterize the convergence in injector nozzles, is the *k-factor*, which is defined as follows in Equation 2.1

$$k\text{-}factor = \frac{D_i - D_o}{10\,\mu\text{m}} \tag{2.1}$$

2.2 Diesel spray fundamentals

The formation of the diesel spray in compression-ignition direct-injection engines, includes very complex processes such as high-velocity jet flow of fuel, liquid droplet break-up, atomization, vaporization of a dense liquid spray and combustion, all of them taking place in a turbulent flow environment and in very small temporal and spacial scales. To ensure a good mixture between the air and the fuel, the spray must penetrate into the combustion chamber and atomize [11–14], which makes the fundamental understanding on the diesel spray formation and behavior, to play a key role in ICEs development. That said, this section aims to summarize some of the well-established knowledge on diesel spray formation to be focused on the presentation of fundamental concepts that are around that process.

2.2.1 Air-fuel mixture formation

The diesel spray structure is normally defined from the outlet of the nozzle orifice, where it is delivered at its determined ρ_f, T_f and $p_f \approx p_{rail}$. Once the fuel is injected into the combustion chamber, which has a turbulent and dense environment at high temperature, there are different processes such as spray atomization, air entrainment, and heat transfer between the environment and droplets (which leads to fuel evaporation), that take place. The first millimeters are characterized by the liquid core, where the intact surface does not present droplet formation. Subsequently, primary atomization occurs, that is what happens with a disintegration of the liquid core and the formation of liquid structures of different sizes, combined with the perturbations deriving from the turbulent flow, the core surface tension and inertial forces. Four different well-defined regimes of liquid core break-up are generally described in literature [15, 16], mentioned here in jet velocity incremental order as shown in Figure 2.8: the Rayleigh mechanism, which happens at low velocities (around 10 m/s) and is caused mainly by axisymetrical oscillations on the liquid core due to the surface tension; first wind-induced regime, whose atomization is faster and the jet adquires a helicoidal shape; second wind-induced regime, where the relative velocity is higher and waves on the liquid surface are quite small, reducing the droplets diameter; and atomization regime, where droplets are formed in the immediate proximity of the orifice and their size is much smaller than the orifice diameter. As the figure shows, an incomplete atomization still shows a larger liquid core and the spray is not as wide as in a full-atomization regime. State-of-the-art engine technologies usually work with sprays in the full-atomization regime where, due to the high jet velocities, other mechanisms such as cavitation and turbulence affect the spray formation.

Gas entrainment into the spray is also produced by this atomization regime: the droplets are so small and the process is so fast that the spray entrains the surrounding air, taking place a momentum transfer between entrained air and the spray. Both gas and fuel droplets interact while travel together penetrating into the combustion chamber, and then, air enters into the spray transferring momentum. Another process that takes part of the air-fuel mixture formation, happens due to the difference of temperature between the gas and the fuel, generating heat transfer from the first to the second one between their interface. Due to this evaporation, droplet diameter is reduced until it is entirely transformed into a gas. Rather than the heat released by the combustion, it is the energy (in terms of enthalpy) of the entrained air that is provided to the droplets what drives the evaporation

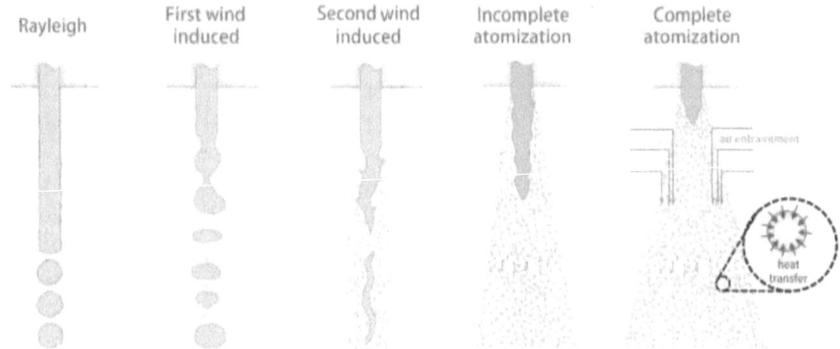

Fig. 2.8: Liquid core brake-up regimes. Last figure is used to illustrate subprocesses of the air-fuel mixture formation. Adaptation made from [17].

rate and that controls the distance from the nozzle where full-evaporation is fulfilled and the spray continues penetrating in fully gaseous state.

2.2.2 The in-chamber diesel combustion

Ignition in compression-ignition engines or autoignition is given by chemical reactions that are triggered by specific conditions of the mixture temperature and air-fuel ratio, which can be locally reached first in particular regions. Regarding the autoignition evolution, Higgins et al. [18] presented one of the most accepted descriptions: First, fuel is atomized and heat exchange with droplets reduce gas temperature in the spray region. This temperature is increased as the spray goes downstream into the chamber, reaching autoignition conditions. At this moment and under oxygen presence, the two stages below, are the following processes:

- **Start of cool-flames (SoCF):** This stage is characterized by a low heat release and intensity of luminescence, and a rich air-fuel ratio. Here, a series of precursor reactions take place, consuming the premixed fuel and increasing chamber pressure and temperature which promotes further chemical reactions [18–21].
- **Second-stage ignition (SSI):** The heat released by the previous stage promotes temperature rise and consequently, the disassociation of the hydrogen peroxide. This process produces the OH radical whose decaying to ground state releases a high level of chemiluminescence. This

stage is characterized by a significant liberation of heat with a steep and localized peak, which is an important source of NOx formation [18, 22, 23].

The apparent heat release rate (also known as AHRR) is one of the most extended means to stablish the temporal evolution of the combustion in diesel engines. It is commonly calculated from the pressure measured in the cylinder [24, 25]. The management of combustion in this type of engines is done by fuel delivery [26–28]. An example of that, extracted from the work of García-Oliver [26] is depicted in Figure 2.9. The ignition delay is defined as the period from the start of the injection up to the instant where induced physical conditions in the air-fuel mixture are compliant to start the combustion. This delay has been widely studied and has shown to be highly dependent on numerous parameters such as ambient temperature and density [12, 23, 29], nozzle diameter [12, 30], rail pressure [21, 22], etc.

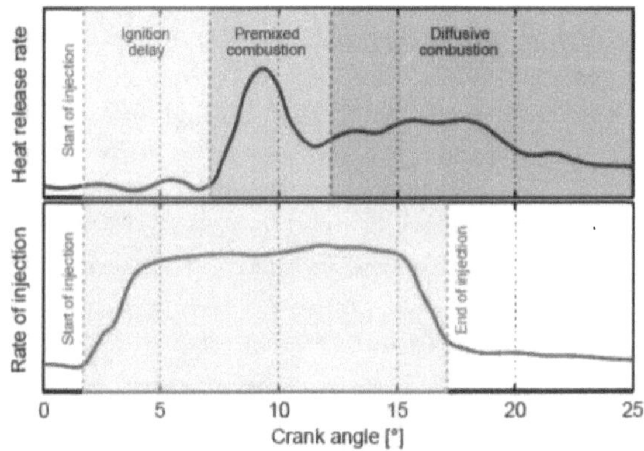

Fig. 2.9: Evolution with time of both apparent heat release rate and mass flow injection rate. Adaptation made from [26].

Once there are regions of the previously injected fuel that are in flammability conditions, it burns quickly, giving place to the first phase of the combustion exothermic reaction where all the fuel premixed before the start of combustion is consumed. This stage is known as premixed combustion and it is characterized by a precipitous heat release increment. After this, the combustion process is driven by the fuel-air mixing that takes place by means of molecular interdiffusion. As a result, the speed of the flame is limited by the rate of mixing. Similarly, this different type of combustion, referred to as diffusion combustion, makes possible a greater uniformity of heat distribution as seen in Figure 2.9 and lower temperature gradients.

This diffusion combustion starts losing stability until its end when the injection is cut. However, as long as fuel and oxygen are supplied to the flame, the flame front is self-sustained [17, 31] and, if injection duration allows it, the flame can achieve a quasi-stationary structure. One of the most generally accepted models for diffusion flames has been presented by the comprehensive study of Dec [32] and is shown in Figure 2.10. Each region is significantly differenced in terms of temperatures and chemical composition. Due to the high jet speed at the nozzle outlet, the flame is 'lifted' a determined distance (known as lift-off length or LoL) in respect of the injector tip, having a first liquid and non-reacting zone.

Starting from this LoL, the flame is formed by the products from a partial combustion (soot and unburnt hydrocarbons) and it is surrounded by a thin reaction layer (whose width has been determined by Dec and Coy [31] to be in the order of 120 µm), where the oxygen entrained by the spray and the gases are mixed and produce CO_2 and H_2O. Due to the high temperatures of this surface, most of the NO_x are formed there. On the tip of this region, a flame vortex head is formed, which has the highest temperature and soot concentration. All the oxygen entrained along the lift-off length reacts in the premixed flame zone, that presents a very rich mixture, therefore, the products (mostly carbon monoxide and unburned polyaromatic hydrocarbons) serve as fuel for the diffusive reaction and constitute the basis for the consequent formation of soot. To summarize, the diesel is supplied into the chamber in form of a jet that mixes with the hot ambient air providing the reactants for a premixed combustion that, due to its intrinsic richness, produces the elements that afterwards will be the fuel for a diffusive flame [33, 34].

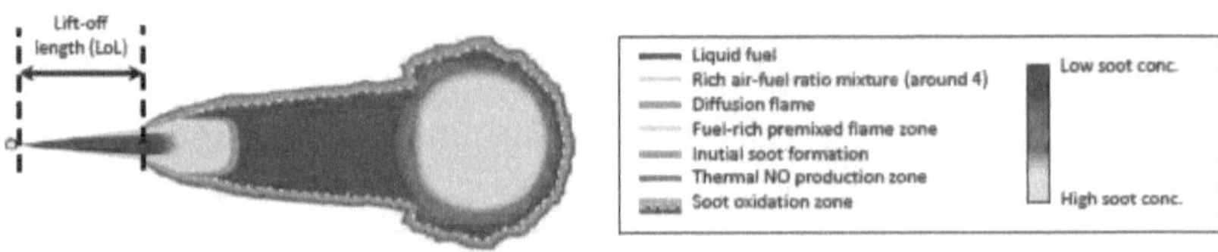

Fig. 2.10: Dec [32] model for a quasy-stationary diffusive flame

2.2.3 Behavior of the transient diesel spray characteristics

Spray penetration is defined as the distance that a spray travels into its medium. Due to its direct relationship with the spray distribution within the chamber and with its suitability as quality indicator of air-fuel mixture and evaporation, it is one of the most widely studied metrics in spray diagnostics [20, 35–39]. Figure 2.11 shows an example of the penetration behavior of a diesel spray in its two vapor and liquid phases, while it goes through a non-reactive atmosphere (no combustion-induced perturbations). At a first moment after start of injection it can be seen how both phases penetrate similarly when evaporation is still negligible. Once the energy of the entrained air is enough to reach full-evaporation, the liquid phase stabilizes around a certain value referred to as liquid length (LL) in literature. On the other hand, the penetration of the vapor phase of the spray remains growing, starting with a high rate and then getting slower with a trend that is proportional to the squared root of time [40–42]. Both phases of penetration are differently affected by operating conditions and nozzle dimensions [42, 43].

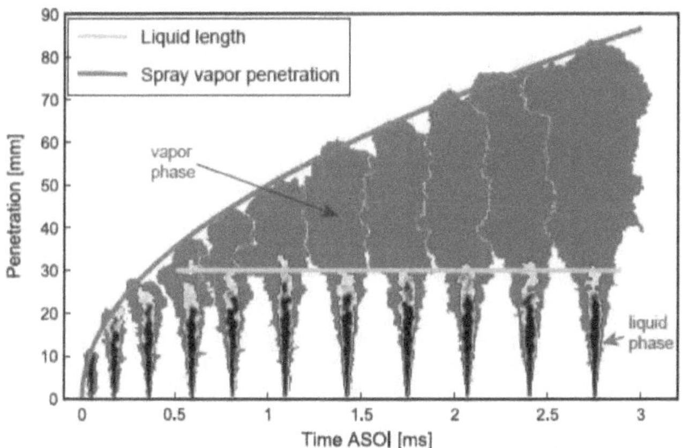

Fig. 2.11: Vapor and liquid phases of a diesel spray penetrating into an inert atmosphere [17].

On the other hand, the spray angle is usually defined as the one that is formed by two lines that coincide in the nozzle outlet and that follow two opposite edges of the spray contour. After a first transient stage, the angle is known to remain constant with time [44, 45]. Nevertheless, its definition is not universal and different criteria can be found in literature to determine its value, being affected not only by its geometrical definition to fit the contour to two lines; but also by the chosen optical technique or the processing methodology. However, several authors agree on its strong dependency on the densities ratio between ambient and fuel [5, 46, 47] and the nozzle geometry [5, 44, 48].

2.3 Impinging diesel sprays

Even when the analysis of free-jet injection and combustion has proven to be exceptionally useful to reach a better understanding on the functioning of injection systems in engine applications and, for the development of accurate empiric and CFD models, it is still a simplistic model of the injection process into a combustion chamber of a reciprocating engine. This is even more important in a world where engine downsizing is a trend and the production of smaller powerplants is more present in the market. Spray-wall interaction has a fundamental role in fuel atomization, mixing and in both combustion behavior and pollutant emissions formation, that are summed up in the scheme shown in Figure 2.12. On the one hand, the incidence and accumulation of fuel in the cylinder walls can lead to the formation of a fuel film that worsens combustion, promotes the emission of carbon monoxide and unburned hydrocarbons and involves energy losses due to the increase of heat transfer. This heat flux from the flame to the wall is also expected to reduce the spray temperature which could delay the ignition process. On the other side, the impact of the spray with a surface gives place to a secondary atomization produced by the impact and also to a larger exposure of the jet surface to the chamber gases, improving the air-fuel mixing which is well-known to enhance combustion. All those elusive effects on internal combustion engines performance, together with its broad technical applications in other fields such as spray-induced cooling, painting and prevention of solid depositions; make spray-wall interaction to remain nowadays as an active area of research.

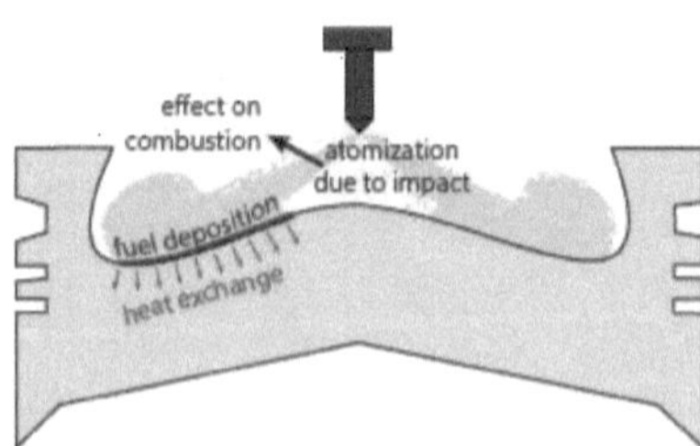

Fig. 2.12: Schematic of the spray-wall interaction process in an engine piston.

2.3.1 Overview on droplet-wall interaction mechanisms

A key way to understand the likelihood of a spray to form a fuel film onto the piston-chamber walls or to enhance break-up of droplets and promote evaporation, is given by the dimensionless Weber number *We*, which is defined as $We = \rho \cdot u^2 \cdot D \cdot \sigma^{-1}$, where ρ is the density of the droplet, u is its impact velocity, D is its characteristic diameter and σ its superficial tension. Several authors have proposed models for the possible regimes of the impact behavior from the study of impingement drops. A map of various collision regimes droplet-wall is shown in Figure 2.13, based on the one presented by Bai and Gosman [49], while similar diagrams have been designed by other authors [50, 51]. In addition to the effect of *We*, the wall temperature has an influence in the manifestation of certain regimes where, accordingly to the plot, T_b is the liquid boiling temperature, T_{PA} is the temperature below which adhesion occurs at low impact energy, T_N is the Nakayama temperature at which a droplet reaches its maximum evaporation rate. T_{PR} is the temperature above which bounce happens at low impact energy and T_{leid} is the Leidenfrost temperature or the point of steady boiling where a vapor layer is formed between the wall and the drop.

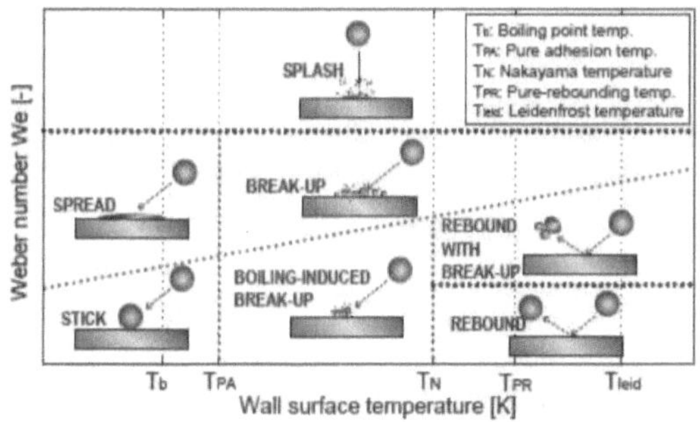

Fig. 2.13: Schematic map of different drop-wall collision regimes proposed by Bai and Gosman [49].

Despite the good qualitative description of *We* and heat induced processes offered by diagrams like Figure 2.13, a fundamental definition of the different regimes is quite complex because of various phenomena observed at cold impact must be addressed within each boiling regime, as a complexity added by surface temperature. Although these heat transfer regimes for impinging droplets-sprays are universal, e.g. Panão and Moreira [52] and Toda [53], collision conditions affect the critical points of the boiling curves. Last decade has given place to extensive investigations on boundary temperatures and their influencing parameters, with valuable and interesting but not always conclusive results: Leidenfrost temperature behavior with a rise on collision velocity can be found to either increase, as done by several authors as Chaves et al. [54] and Yao and Cai [55], or to decrease [56] as a result of a variation of the convection behavior. Similarly inconclusive, several researches [57, 58] report negligible effect of droplet diameter on boundary temperatures while, for instance, Shi et al. [59] observed an increment of T_{leid} with droplet size that was explained by the force balance between dynamic pressure of droplets and the vapor pressure of the vapor layer. For sprays, boundary temperature estimation is even more complicated to assess in a large extent, due to non-uniform distribution of liquid mass over the impacted area [52], making the application of similarity rules between the heat transfer over the entire impacted area and a single model not possible and not expected to hold.

Nowadays, few studies reported in the literature [60–62] are still at early stages, and it is undeniable that multiple drop interaction phenomena are one of the most prominent open topics for research. However, accurate integration between both interaction phenomena of droplets and spray impingement with walls is still far from being successfully achieved [60, 63], which lets open the door to go forward with more realistic approaches such as impinging sprays characterization.

2.3.2 Spray-wall interaction at inert atmospheres

In order to attend the ever-growing need for cleaner engines, it is important to extend the knowledge from the free-spray perspective to take into account the interaction with the combustion chamber boundaries. Likewise, a more approximated approach from the droplet-wall impingement analysis, is made from spray-wall interaction studies. In particular, spray-wall contact has a huge influence on powertrain emissions and engine performance [64–69]. In low pressure systems as, for instance, PFI (Port Fuel Injection) spark ignited engines, vaporization is not fully achieved in the port and drops are deposited on the chamber walls, which is referred to as wall-wetting in literature [70]. This phenomenon is also present during cold-start conditions in direct injection diesel engines. Reitz and Duraisamy [71] have shown how efficiency is shortened and pollutant emissions are considerably intensified particularly in low-temperature concepts such as RCCI and PCCI (Reactivity Controlled Compression Ignition and Premixed Charge Compression Ignition respectively). Several works have tried to define the macroscopic characteristics of an impinging spray and its different regions with a general agreement. An adaptation of some of those conceptual models [64, 72–74] is shown in Figure 2.14.

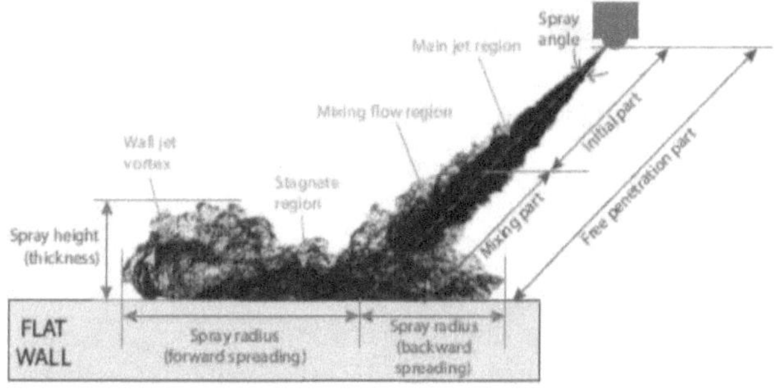

Fig. 2.14: Model of a diesel impinging spray against a flat wall.

Under non-reactive conditions, effects of injection pressure and wall temperature have been observed in [75] for a GDi single-hole injector via optical techniques to visualize both liquid and vapor phases. Authors observed that, at the same pressure, radial spreading is affected by wall temperature but could not find a direct correlation for maximum thickness of the spray along the wall. Nevertheless, for a fixed wall temperature, they found a linear spreading growth with pressure injection and a similar proportion for spray spreading. Yu et al. [76, 77] experimentally tested different fuels and and injection strategies, finding that gas-entrainment and spray momentum have a great effect on wall film formation and spray spreading, whereas the viscosity and surface tension directly affected spray thickness along the wall. PDA (Phase Doppler Anemometry) and high-speed imaging methods were used by Montorsi et al. [78] who employed three coaxial thermocouples in the wall and set the chamber to conditions corresponding to those found during the compression stroke in a heavy duty diesel engine. Heat flux through the wall was calculated using a 1-D transient heat model, along with simulations on STAR-CD, KIVA-3V, concluding that SWI is sensitive to ambient pressure and wall temperature; therefore both heat-release rate and emission formation were affected. A similar experimental approach was used by Meingast et al. [66] in a heated wall, obtaining that factors that promote smaller droplet sizes at the moment of impingement, such as higher injection pressures and wall temperatures or larger distances between injector and wall lead to a better gas-liquid mixing and a lower wall heat flux. Accordingly to Zhu et al. [79], who performed a study on non-vaporizing sprays at high-density and high-injection pressures with an impingement angle of 60°, the wall-impinged expanding spray evolution is divided into four different stages: rapidly decelerated stage, slowly decreasing rate stage, relatively constant rate stage, and expanding termination stage.

Gas-fuel mixing process has been analyzed by Bruneaux [80] via LIEF (Laser Induced Exciplex Fluorescence), comparing both perpendicular spray-wall impact and free-jet situations. This investigation showed that mixing is reduced compared to the free-jet at the center impingement region near the spray axis, while in the region where the spray moves along the wall, mixing is increased again and that changes the contribution of both effects depending on the injection pressure, due to the change on impingement moment, turbulence, etc. Zhao et al. [81] studied SWI using both a constant volume combustion vessel and a RANS based methodology for simulations, observing an enhanced spray momentum at the impinging wall under low ambient densities or high injection pressure.

2.3.3 Spray-wall interaction at reacting atmospheres

Reacting sprays and flame interaction with combustion chamber walls is also inevitable and with a large impact on efficiency and emissions control. As the main target of impinging diesel sprays is the piston bowl, its design has been piece of study in order to optimize engines performance [69, 82–85]. Nevertheless, due to the fixed nozzle orientation in engine injectors and injections taking place at different crank angles (different conditions and loads, multi-injection strategies, etc.) complicate to achieve an overall optimization [86, 87]. Not only piston diameters and bore-to-stroke ratios [82, 88] have shown to impact heat losses to metallic walls, but also variations respect to convencional piston bowl geometries are used to isolate sprays (i.e. non-axisymetric spray-confining [69]) or to split the spray between the bowl and the squish region ('stepped-lip' bowls [85, 88]) and affect the fuel spreading and soot formation. However, many researches still agree on the need of more basic understanding on spray-wall interaction and further improvements in diesel engine CFD modelling, to predict tendencies on engines performance [83, 88].

More simplified and fundamental studies have been carried out by researchers in order to isolate variations. A two-dimensional piston cavity and a single-hole injector whose axis was set at 13.5° from the 'cylinder head' were used by Li et al. [89] in a constant volume combustion vessel in order to analyze the combustion process and the flame structure via two-color pyrometry. A higher concentration of soot was found to be located in the head vortex region, while in the case of free spray flame, it is close to the flame tip. Their results with the same injector set at 90° respect to the flat piston head applying OH* chemiluminescence and natural luminosity [90] revealed that, in comparison with a free spray flame, flat-wall impingement causes a deterioration of diesel combustion when the liquid phase-wall interaction takes place and that it can be overcame with an appropriate impinging distance longer than liquid-phase penetration. Similarly, this study found that soot formation decreases with a rise in injection pressure.Wang et al. [91] agree on the finding of liquid impingement being responsible for a worse soot level control respect to a free flame. Combustion structure of free and wall-impinging diesel jets has been also studied by Bruneaux [65] in low-temperature conditions applying simultaneously planar LIF of the OH* radicals with excitation near 280 nm and LIF with excitation at 355 nm in order to spatially identify the fluorescence of formaldehyde and poly-aromatic hydrocarbons (PAHs). The combustion structure obtained by this piece of work is shown in Figure 2.15. As seen above, the Dec model [32] is quite similar to this one in terms of the free-jet situation; the PAH and soot cloud was extended up to the jet tip, and just a thin layer of OH LIF at the spray edge is observed. However, it is important to point out that Bruneaux [65] conceptual model is valid for the moderate-soot conditions of that work (T_{amb} = 800 K; ρ_{amb} = 25 kg m^{-3}; xO_2 = 0.21) in contrast to the model of Dec where the ambient temperature in which it was conceived was 1000 K. In Figure 2.15, the central part of the combustion structure is similar in both free-jet and SWI situations, but broader regions of OH are observed at the jet tip in the spray-wall case respect to the thin lines in the free jet configuration, as a result of improved mixing by the jet-wall vortex in accordance with the conclusion in [80] about fuel-air mixture formation.

In spite of the existance of research on soot production under SWI conditions, some of their results are still ambiguous and a conclusive understanding might miss. For instance, the mentioned study of Wang et al. [91] presents how the insertion of a wall increases soot formation, specially when it is placed near to the injector. On the other hand, Pickett and Lopez [92] found in the use of SWI respect to free-jet, the potential to reduce or even eliminate soot formation, by means of the increased fuel-air mixing rate and the wall-jet-cooling effect provoked by SWI. That last experiment consisted in the use of a flat wall and also a rectangular box-shaped wall to confine the colliding jet and simulate the jet-jet interaction of adjacent jets, with the single-hole injector used in the test. Nevertheless, they found that jet confinement causes combustion gases to be redirected towards the incoming jet, shortening the lift-off length and rising soot formation, evidencing the higher level of complexity in realistic diesel engines configurations. Maes et

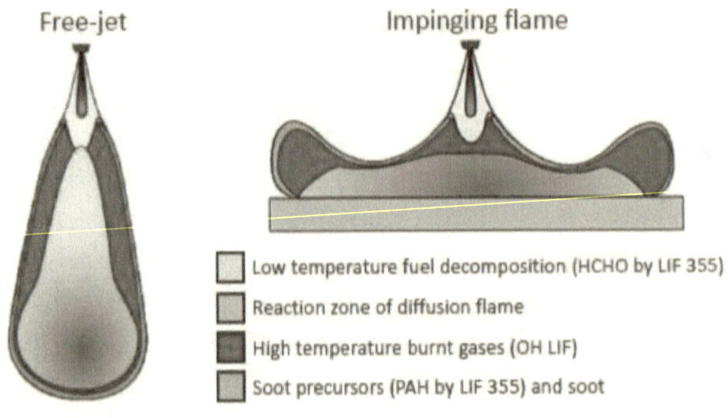

Fig. 2.15: Conceptual models of diesel jet combustion for both a free jet and an impinging jet during the stabilized diffusion combustion phase presented by Bruneaux [65].

al. [27] performed another study with alternative wall shapes (a flat one, a 2-D confined wall and an axisymmetric confined wall) obtaining wall temperature and high-speed pressure-based heat release data wall temperature. They do corroborate an improved mixing due to SWI and a faster increase of the heat release rate, and this effect is intensified at shorter injector-wall distances. Additionally, this investigation found a reduction on soot formation when the flame is dispersed, cooled and quenched by the wall surface and when the wall surface is moved closer to the injector orifice.

References

[1] Halderman, J.D. and Linder, J. *Automotive Fuel and Emissions Control Systems*. Prentice Hall, 2011 (cited on page 14).

[2] Jaramillo Císcar, David. "Estudio experimental y computacional del proceso de inyección diésel mediante un código CFD con malla adaptativa". PhD book. Universitat Politècnica de València, 2017 (cited on page 16).

[3] Manin, Julien. "Analysis of mixing processes in liquid and vaporized diesel sprays through LIF and Rayleigh scattering measurements". PhD book. Valencia: E.T.S. Ingenieros Industriales. Universidad Politécnica de Valencia, 2011 (cited on page 17).

[4] Salvador, Francisco Javier, Plazas, Alejandro Hernan, Gimeno, Jaime, and Carreres, Marcos. "Complete modelling of a piezo actuator last-generation injector for diesel injection systems". In: *International Journal of Engine Research* 15.1 (2014), pp. 3–19 (cited on page 19).

[5] Mitroglou, Nicholas, Gavaises, Manolis, Nouri, J M, and Arcoumanis, Constantine. "Cavitation Inside Enlarged and Real-Size Fully Transparent Injector Nozzles and Its Effect on Near Nozzle Spray Formation". In: *DIPSI Workshop 2011 on Droplet Impact Phenomena and Spray Investigation, Bergamo, Italy.* (2011) (cited on pages 20, 27).

[6] Schmidt, David P. and Corradini, M L. "The internal flow of Diesel fuel injector nozzles: a review". In: *International Journal of Engine Research* 2.6 (2001), pp. 1–22 (cited on page 20).

[7] Marti-Aldaravi, Pedro, Saha, Kaushik, Gimeno, Jaime, and Som, Sibendu. "Numerical Simulation of a Direct-Acting Piezoelectric Prototype Injector Nozzle Flow for Partial Needle Lifts". In: *SAE Technical Papers* 2017-24-01 (2017) (cited on page 20).

[8] Fitzgerald, Russell P, Vecchia, Giovanni Della, Peraza, Jesús E, and Martin, Glen C. "Features of Internal Flow and Spray for a Multi-Hole Transparent Diesel Fuel Injector Tip". In: *ILASS Europe 2019, 29th Conference on Liquid Atomization and Spray Systems* September (2019), pp. 2–4 (cited on pages 20, 21).

[9] Payri, Raul, Gimeno, Jaime, Martí-Aldaraví, Pedro, and Venegas, Oscar. "Study of the influence of internal flow on the spray behavior under cavitating conditions using a transparent nozzle". In: *ICLASS 2012, 12th Triennial International Conference on Liquid Atomization and Spray Systems.* 2012, pp. 1–8 (cited on page 21).

[10] Roth, H. et al. "Effect of multi-injection strategy on cavitation development in diesel injector nozzle holes". In: *SAE transactions* 114.3 (2005), pp. 1029–1045 (cited on page 21).

[11] Soid, S N and Zainal, Z A. "Spray and combustion characterization for internal combustion engines using optical measuring techniques - a review". In: *Energy* 36.2 (2011), pp. 724–741 (cited on page 21).

[12] Payri, Raul, Viera, Juan Pablo, Gopalakrishnan, Venkatesh, and Szymkowicz, Patrick G. "The effect of nozzle geometry over ignition delay and flame lift-off of reacting direct-injection sprays for three different fuels". In: *Fuel* 199 (2017), pp. 76–90 (cited on pages 21, 24).

[13] Neal, Nicholas and Rothamer, David. "Measurement and characterization of fully transient diesel fuel jet processes in an optical engine with production injectors". In: *Experiments in Fluids* 57.10 (2016), p. 155 (cited on page 21).

[14] Hwang, Joonsik, Park, Youngsoo, Kim, Kihyun, Lee, Jinwoo, and Bae, Choongsik. "Improvement of diesel combustion with multiple injections at cold condition in a constant volume combustion chamber". In: *Fuel* 197 (2017), pp. 528–540 (cited on page 21).

[15] Reitz, Rolf D. "Atomisation and other breakup regimes of a liquid jet". PhD book. Ph.D. book, Princeton University, 1978 (cited on page 22).

[16] Kolev, Nikolay Ivanov. "Liquid and gas jet disintegration". In: *Multiphase Flow Dynamics 2: Mechanical Interactions.* Berlin, Heidelberg: Springer Berlin Heidelberg, 2012, pp. 287–310 (cited on page 22).

[17] Bardi, Michele. "Partial needle lift and injection rate shape effect on the formation and combustion of the Diesel spray". PhD book. Valencia (Spain): Universitat Politècnica de València, 2014 (cited on pages 23, 25, 26).

[18] Higgins, Brian, Siebers, Dennis L, and Aradi, Allen. "Diesel-Spray Ignition and Premixed-Burn Behavior". In: *SAE Technical Paper 2000-01-0940* (2000) (cited on pages 23, 24).

[19] Pickett, Lyle M, Siebers, Dennis L, and Idicheria, Cherian A. "Relationship Between Ignition Processes and the Lift-Off Length of Diesel Fuel Jets". In: *SAE Paper 2005-01-3843* 724 (2005) (cited on page 23).

[20] Gimeno, Jaime, Martí-Aldaraví, Pedro, Carreres, Marcos, and Peraza, Jesús E. "Effect of the nozzle holder on injected fuel temperature for experimental test rigs and its influence on diesel sprays". In: *International Journal of Engine Research* 19.3 (2018), pp. 374–389 (cited on pages 23, 26).

[21] Payri, Raul, Salvador, Francisco Javier, Manin, Julien, and Viera, Alberto. "Diesel ignition delay and lift-off length through different methodologies using a multi-hole injector". In: *Applied Energy* 162 (2016), pp. 541–550 (cited on pages 23, 24).

[22] Benajes, Jesus, Payri, Raul, Bardi, Michele, and Martí-aldaraví, Pedro. "Experimental characterization of diesel ignition and lift-off length using a single-hole ECN injector". In: *Applied Thermal Engineering* 58.1-2 (2013), pp. 554–563 (cited on page 24).

[23] Payri, Raul, Salvador, Francisco Javier, Gimeno, Jaime, and Peraza, Jesús E. "Experimental study of the injection conditions influence over n-dodecane and diesel sprays with two ECN single-hole nozzles. Part II: Reactive atmosphere". In: *Energy Conversion and Management* 126 (2016), pp. 1157–1167 (cited on page 24).

[24] Mungal, M G and Mun, L. "Effects of Heat Release and Buoyancy on Flow Structure and Entrainment in Turbulent Nonpremixed Flames". In: *Combustion and Flame* 126 (2001), pp. 1402–1420 (cited on page 24).

[25] Jorques Moreno, Carlos and Stenlaas, Ola. "Influence of Small Pilot on Main Injection in a Heavy-Duty Diesel Engine". In: *SAE Technical Paper 2017-01-0708* (2017) (cited on page 24).

[26] García-Oliver, José María. "Aportaciones al estudio del proceso de combustión turbulenta de chorros en motores Diesel de inyección directa". PhD book. Valencia: E.T.S. Ingenieros Industriales. Universitat Politècnica de València, 2004 (cited on page 24).

[27] Maes, Noud, Hooglugt, Mark, Dam, Nico, Somers, Bart, and Hardy, Gilles. "On the influence of wall distance and geometry for high-pressure n-dodecane spray flames in a constant-volume chamber". In: *International Journal of Engine Research* (2019) (cited on pages 24, 32).

[28] Lillo, Peter M, Pickett, Lyle M, Persson, Helena, Andersson, Öivind, and Kook, Sanghoon. "Diesel Spray Ignition Detection and Spatial/Temporal Correction". In: *SAE Technical Paper 2012-01-1239* (2012) (cited on page 24).

[29] Payri, Raul, Garcia-Oliver, Jose Maria, Bardi, Michele, and Manin, Julien. "Fuel temperature influence on diesel sprays in inert and reacting conditions". In: *Applied Thermal Engineering* 35 (2012), pp. 185–195 (cited on page 24).

[30] Fitzgerald, Russell P, Svensson, Kenth, Martin, Glen, Qi, Yongli, and Koci, Chad. "Early Investigation of Ducted Fuel Injection for Reducing Soot in Mixing-Controlled Diesel Flames". In: *SAE Technical Paper 2018-01-0238* (2018), pp. 1–17 (cited on page 24).

[31] Dec, John E and Coy, Edward B. "OH Radical Imaging in a DI Diesel Engine and the Structure of the Early Diffusion Flame". In: 412 (1996) (cited on page 25).

[32] Dec, John E. "A Conceptual Model of DI Diesel Combustion Based on Laser-Sheet Imaging". In: *SAE Technical Paper 970873* (1997) (cited on pages 25, 32).

[33] Flynn, P et al. "Diesel combustion: an integrated view combining laser diagnostics, chemical kinetics, and empirical validation". In: *SAE Paper 1999-01-0509* 724 (1999) (cited on page 25).

[34] Huang, Sheng et al. "Visualization research on spray atomization, evaporation and combustion processes of ethanol-diesel blend under LTC conditions". In: *Energy Conversion and Management* 106 (2015), pp. 911–920 (cited on page 25).

[35] Araneo, Lucio, Coghe, Aldo, Brunello, G, and Cossali, Gianpietro E. "Experimental Investigation of gas density effects on diesel spray penetration and entrainment". In: *SAE Paper 1999-01-0525* (1999) (cited on page 26).

[36] Pickett, Lyle M, Genzale, Caroline L, Manin, Julien, Malbec, Louis-Marie, and Hermant, Laurent. "Measurement Uncertainty of Liquid Penetration in Evaporating Diesel Sprays". In: *ILASS Americas 23rd Annual Conference on Liquid Atomization and Spray Systems.* Ventura, CA (USA): ILASS-Americas, 2011 (cited on page 26).

[37] Westlye, Fredrik R et al. "Penetration and combustion characterization of cavitating and non-cavitating fuel injectors under diesel engine conditions". In: *SAE Technical Paper 2016-01-0860* (2016), p. 15 (cited on page 26).

[38] Payri, Raul, Salvador, Francisco Javier, Marti-Aldaravi, Pedro, and Vaquerizo, Daniel. "ECN Spray G external spray visualization and spray collapse description through penetration and morphology analysis". In: *Applied Thermal Engineering* 112 (2017), pp. 304–316 (cited on page 26).

[39] Desantes, Jose Maria, Payri, Raul, Salvador, Francisco Javier, and Gimeno, Jaime. "Prediction of Spray Penetration by Means of Spray Momentum Flux". In: *SAE Technical Paper 2006-01-1387* (2006) (cited on page 26).

[40] Naber, Jeffrey D and Siebers, Dennis L. "Effects of Gas Density and Vaporization on Penetration and Dispersion of Diesel Sprays". In: *SAE Paper 960034* (1996) (cited on page 26).

[41] Desantes, Jose Maria, Payri, Raul, Salvador, Francisco Javier, and Gil, Antonio. "Development and validation of a theoretical model for diesel spray penetration". In: *Fuel* 85.7-8 (2006), pp. 910–917 (cited on page 26).

[42] Gimeno, Jaime, Bracho, Gabriela, Martí-Aldaraví, Pedro, and Peraza, Jesús E. "Experimental study of the injection conditions influence over n-dodecane and diesel sprays with two ECN single-hole nozzles. Part I: Inert atmosphere". In: *Energy Conversion and Management* 126 (2016), pp. 1146–1156 (cited on page 26).

[43] Kook, Sanghoon and Pickett, Lyle M. "Liquid length and vapor penetration of conventional , Fischer-Tropsch , coal-derived , and surrogate fuel sprays at high-temperature and high-pressure ambient conditions". In: *Fuel* 93 (2012), pp. 539–548 (cited on page 26).

[44] Jung, Yongjin, Manin, Julien, Skeen, Scott A, and Pickett, Lyle M. "Measurement of Liquid and Vapor Penetration of Diesel Sprays with a Variation in Spreading Angle". In: *SAE Technical Paper 2015-01-0946* (2015) (cited on pages 26, 27).

[45] Gimeno, Jaime. "Desarrollo y aplicación de la medida de flujo de cantidad de movimiento de un chorro Diesel". PhD book. E.T.S. Ingenieros Industriales. Universitat Politècnica de València, 2008 (cited on page 26).

[46] Payri, Raul, Salvador, Francisco Javier, Gimeno, Jaime, and Viera, Juan Pablo. "Experimental analysis on the influence of nozzle geometry over the dispersion of liquid n-dodecane sprays". In: *Frontiers in Mechanical Engineering* 1 (2015), pp. 1–10 (cited on page 27).

[47] Giraldo, Jhoan S, Payri, Raul, Marti-Aldaravi, Pedro, and Montiel, Tomas. "Effect of high injection pressures and ambient gas properties over the macroscopic characteristics of the diesel spray on multi-hole nozzles". In: *Atomization and Sprays* 28.12 (2019), pp. 1145–1160 (cited on page 27).

[48] Payri, Raul, Viera, Juan Pablo, Gopalakrishnan, Venkatesh, and Szymkowicz, Patrick G. "The effect of nozzle geometry over the evaporative spray formation for three different fuels". In: *Fuel* 188 (2017), pp. 645–660 (cited on page 27).

[49] Bai, Chengxin and Gosman, A. D. "Development of Methodology for Spray Impingement Simulation". In: 412 (1995) (cited on page 28).

[50] Akhtar, S W and Yule, A J. "Droplet impaction on a heated surface at high Weber numbers". In: *ILASS-Europe, Zurich* September 2001 (2001) (cited on page 28).

[51] Lee, Sang Yong and Ryu, Sung Uk. "Recent progress of spray-wall interaction research". In: *Journal of Mechanical Science and Technology* 20.8 (2006), pp. 1101–1117 (cited on page 28).

[52] Panão, M.R.O. and Moreira, A.L.N. "Thermo- and fluid dynamics characterization of spray cooling with pulsed sprays". In: *Experimental Thermal and Fluid Science* 30 (2005), pp. 79–96 (cited on pages 28, 29).

[53] Toda, Saburo. "A Study of Mist Cooling : 1st Report, Experimental Investigations on Mist Cooling by Mist Flow Sprayed Vertically on Small and Flat Plates Heated at High Temperatures". In: *Transactions of the Japan Society of Mechanical Engineers* 38.307 (1972), pp. 581–588 (cited on page 28).

[54] Chaves, Humberto, Kubitzek, Artur Michael, and Obermeier, Frank. "Dynamic processes occurring during the spreading of thin liquid films produced by drop impact on hot walls". In: *International Journal of Heat and Fluid Flow* 20.5 (1999), pp. 470–476 (cited on page 29).

[55] Yao, Shi-Chune and Cai, Kang Yuan. "The dynamics and leidenfrost temperature of drops impacting on a hot surface at small angles". In: *Experimental Thermal and Fluid Science* 1.4 (1988), pp. 363–371 (cited on page 29).

[56] Celata, G.P., Cumo, M., Mariani, A., and Zummo, Giuseppe. "Visualization of the impact of water drops on a hot surface: Effect of drop velocity and surface inclination". In: *Heat and Mass Transfer* 42 (2006), pp. 885–890 (cited on page 29).

[57] Labeish, V.G. "Thermohydrodynamic study of a drop impact against a heated surface". In: *Experimental Thermal and Fluid Science* 8.3 (1994), pp. 181–194 (cited on page 29).

[58] Bernardin, J. D. and Mudawar, I. "The Leidenfrost Point: Experimental Study and Assessment of Existing Models". In: *Journal of Heat Transfer* 121.4 (1999), pp. 894–903 (cited on page 29).

[59] Shi, M.H., Bai, T.C., and Yu, J. "Dynamic behavior and heat transfer of a liquid droplet impinging on a solid surface". In: *Experimental Thermal and Fluid Science* 6.2 (1993), pp. 202–207 (cited on page 29).

[60] Moreira, A.L.N. and Moita, A.s. "Droplet-Wall Interactions". In: 2011, pp. 183–197 (cited on page 29).

[61] Zhang, Fan, Ma, T Y, Zhang, F, Liu, H F, and Yao, M F. "Modeling of droplet / wall interaction based on SPH method". In: *International Journal of Heat and Mass Transfer* 105.February (2017), pp. 296–304 (cited on page 29).

[62] Li, Chao, Wu, Guanjie, Li, Mengzhe, Hu, Chunbo, and Wei, Jinjia. "A heat transfer model for aluminum droplet/wall impact". In: *Aerospace Science and Technology* 97 (2020), p. 105639 (cited on page 29).

[63] Moreira, A.L.N., Moita, A S, and Panão, Miguel R. O. "Advances and challenges in explaining fuel spray impingement : How much of single droplet impact research is useful ?" In: *Progress in Energy and Combustion Science* 36 (2010), pp. 554–580 (cited on page 29).

[64] Stanton, Donald W. and Rutland, Christopher J. "Multi-dimensional modeling of thin liquid films and spray-wall interactions resulting from impinging sprays". In: *International Journal of Heat and Mass Transfer* 41.20 (1998), pp. 3037–3054 (cited on pages 29, 30).

[65] Bruneaux, Gilles. "Combustion structure of free and wall-impinging diesel jets by simultaneous laser-induced fluorescence of formaldehyde, poly-aromatic hydrocarbons, and hydroxides". In: *International Journal of Engine Research* 9.3 (2008), pp. 249–265 (cited on pages 29, 32, 33).

[66] Meingast, Ulrich, Staudt, Michael, Reichelt, Lars, and Renz, Ulrich. "Analysis of Spray / Wall Interaction Under Diesel Engine Conditions". In: *SAE Technical Paper 2000-01-0272* 724 (2000), pp. 1–15 (cited on pages 29, 30).

[67] Cossali, G E, Coghe, A, and Marengo, M. "The impact of a single drop on a wetted solid surface". In: *Experiments in Fluids* 22 (1997), pp. 463–472 (cited on page 29).

[68] Payri, Raul, Gimeno, Jaime, Peraza, Jesús E., and Bazyn, Tim. "Spray / wall interaction analysis on an ECN single-hole injector at diesel-like conditions through Schlieren visualization". In: *Proc. 28th ILASS-Europe, Valencia* September (2017) (cited on page 29).

[69] Dempsey, Adam B., Seiler, Patrick, Svensson, Kenth, and Qi, Yongli. "A Comprehensive Evaluation of Diesel Engine CFD Modeling Predictions Using a Semi-Empirical Soot Model over a Broad Range of Combustion Systems". In: *SAE Int. J. Engines* 11 (2018), pp. 1399–1420 (cited on pages 29, 31).

[70] Stanglmaier, Rudolf H., Li, Jianwen, and Matthews, Ronald D. "The Effect of In-Cylinder Wall Wetting Location on the HC Emissions from SI Engines". In: *SAE Technical Paper 1999-01-0502*. SAE International, 1999 (cited on page 29).

[71] Reitz, Rolf D and Duraisamy, Ganesh. "Review of high ef fi ciency and clean reactivity controlled compression ignition (RCCI) combustion in internal combustion engines". In: *Progress in Energy and Combustion Science* 46 (2015), pp. 12–71 (cited on page 29).

[72] Jia, Ming et al. *Development of a New Spray/Wall Interaction Model for Diesel Spray Under Pcci-Engine Relevant Conditions*. Vol. 24. 1. 2013, pp. 41–80 (cited on page 30).

[73] Arai, Masataka, Amagai, Kenji, Nagataki, Tsubasa, and Okita, Hideki. "Ignition Positions of a Diesel Spray Impinging on an Inclined Wall". In: (2005) (cited on page 30).

[74] Akop, Mohd Zaid, Zama, Yoshio, Furuhata, Tomohiko, and Arai, Masataka. "Characteristics Of Adhesion Diesel Fuel On An Impingement Disk Wall Part 1: Effect Of Impingement Area And Inclination Angle Of Disk". In: *Atomization and Sprays* 23.8 (2013), pp. 725–724 (cited on page 30).

[75] Montanaro, Alessandro, Allocca, Luigi, Meccariello, Giovanni, and Lazzaro, Maurizio. "Schlieren and Mie Scattering Imaging System to Evaluate Liquid and Vapor Contours of a Gasoline Spray Impacting on a Heated Wall". In: *SAE Technical Papers* 2015 (2015) (cited on page 30).

[76] Yu, Hanzhengnan, Liang, Xingyu, Shu, Gequn, Wang, Yuesen, and Zhang, Hongsheng. "Experimental investigation on spray-wall impingement characteristics of n-butanol/diesel blended fuels". In: *Fuel* 182 (2016), pp. 248–258 (cited on page 30).

[77] Yu, Hanzhengnan, Liang, Xingyu, Shu, Gequn, Sun, Xiuxiu, and Zhang, Hongsheng. "Experimental investigation on wall film ratio of diesel, butanol/diesel, DME/diesel and gasoline/diesel blended fuels during the spray wall impingement process". In: *Fuel Processing Technology* 156 (2017), pp. 9–18 (cited on page 30).

[78] Montorsi, Luca, Magnusson, Alf, and Andersson, Sven. "A Numerical and Experimental Study of Diesel Fuel Sprays Impinging on a Temperature Controlled Wall". In: *SAE Technical Paper 2006-01-3333* 724 (2006), pp. 776–790 (cited on page 30).

[79] Zhu, Xiucheng et al. "An Experimental Study of Diesel Spray Impingement on a Flat Plate: Effects of Injection Conditions". In: September (2017) (cited on page 30).

[80] Bruneaux, Gilles. "Mixing Process in High Pressure Diesel Jets by Normalized Laser Induced Exciplex Fluorescence Part II: Wall Impinging Versus Free Jet". In: *SAE Technical Paper 2005-01-2100* c (2005) (cited on pages 31, 32).

[81] Zhao, Le et al. "Evaluation of Diesel Spray-Wall Interaction and Morphology around Impingement Location". In: 2018 (cited on page 31).

[82] Miles, Paul C and Andersson, Öivind. "A review of design considerations for light-duty diesel combustion systems". In: *International Journal of Engine Research* 17.1 (2016), pp. 6–15 (cited on page 31).

[83] Perini, Federico et al. "Piston geometry effects in a light-duty, swirl-supported diesel engine: Flow structure characterization". In: *International Journal of Engine Research* 19.10 (2018), pp. 1079–1098 (cited on page 31).

[84] Chen, Yanlin, Li, Xiangrong, Li, Xiaolun, Zhao, Weihua, and Liu, Fushui. "The wall-flow-guided and interferential interactions of the lateral swirl combustion system for improving the fuel/air mixing and combustion performance in DI diesel engines". In: *Energy* 166 (2019), pp. 690–700 (cited on page 31).

[85] Busch, Stephen et al. "Bowl Geometry Effects on Turbulent Flow Structure in a Direct Injection Diesel Engine". In: *International Powertrains, Fuels & Lubricants Meeting*. SAE International, 2018 (cited on page 31).

[86] O'Connor, Jacqueline and Musculus, Mark. *Post Injections for Soot Reduction in Diesel Engines: A Review of Current Understanding*. 2013 (cited on page 31).

[87] Han, Zhiyu, Uludogan, All, Hampson, Gregory J., and Reitz, Rolf D. "Mechanism of Soot and NOx Emission Reduction Using Multiple-injection in a Diesel Engine". In: *International Congress & Exposition*. SAE International, 1996 (cited on page 31).

[88] XiangRong, Li, WeiHua, Zhao, HaoBu, Gao, and FuShui, Liu. "Fuel and air mixing characteristics of wall-flow-guided combustion systems under a low excess air ratio condition in direct injection diesel engines". In: *Energy* 175 (2019), pp. 554–566 (cited on page 31).

[89] Li, Kuichun et al. "Effect of Spray/Wall Interaction on Diesel Combustion and Soot Formation in Two-Dimensional Piston Cavity". In: *SAE International Journal of Engines* 6.4 (2013), pp. 2061–2071 (cited on page 31).

[90] Li, Kuichun, Nishida, Keiya, Ogata, Youichi, and Shi, Baolu. "Effect of flat-wall impingement on diesel spray combustion". In: *Proceedings of the Institution of Mechanical Engineers, Part D: Journal of Automobile Engineering* 229.5 (2015), pp. 535–549 (cited on page 32).

[91] Wang, Xiangang, Huang, Zuohua, Zhang, Wu, Kuti, Olawole Abiola, and Nishida, Keiya. "Effects of ultra-high injection pressure and micro-hole nozzle on flame structure and soot formation of impinging diesel spray". In: *Applied Energy* 88.5 (2011), pp. 1620–1628 (cited on page 32).

[92] Pickett, Lyle M and Lopez, Jose Javier. "Jet-wall interaction effects on diesel combustion and soot formation". In: 2005.724 (2005) (cited on page 32).

Chapter 3

Experimental apparatus and tools

This chapter describes the experimental tools employed to carry out the experiments and to analyze the extracted data. To begin with, the fuel injection system is detailed, with emphasis in the injector used in this work. The next section will be used to describe the optically accessible high-pressure and high-temperature facility in which the measurements were made. Two following sections are dedicated to present the two wall systems that were devised in order to adapt the test rig to different kind of tests related to spray-wall interaction studies. Lastly, a section about the calculations made for the design of the thermo-regulated steel wall is included. All measuring campaigns were performed at the CMT-Motores Térmicos institute.

3.1 Injection system

The system employed to deliver fuel to the injector is composed by normal production series components that have been adapted to laboratory use. It primarily comprises a high-pressure common-rail unit and the injector. Hereunder, these systems and the fuels used in this work, are described.

3.1.1 High-pressure unit

A scheme of the high-pressure unit is shown in Figure 3.1. Fuel is extracted from a tank and it passes through an air purge device and a particle filter. Then, it is pressurized by a Bosch CP3 pump driven by an electric motor. After that, fuel is delivered to a common-rail provided with a pressure regulator which is controlled with a closed-loop proportional-integral-derivative (PID). Low-pressure fuel returns from the pump and the common-rail and it is sent to a heat exchanger to the beginning of the circuit. This is important, not only to keep under control the experiment in a better extent, but also to avoid abrupt fuel temperature increments due to the high rate of fuel returning to the tank compared to the consumption rate. The common-rail has a 22 cc volume and is connected to the injector by a tube of 200 mm length and an inner diameter of 3 mm.

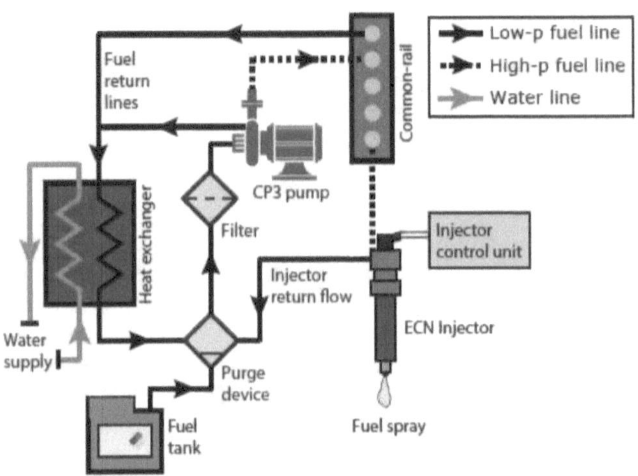

Fig. 3.1: Scheme of the high-pressure unit.

3.1.2 ECN injector: 'Spray D'

The injector used in this work is a single-hole Bosch 3-22 type that is part of the Engine Combustion Network (ECN) dataset, known as Spray D. Its nozzle is based on Spray A, a target ECN single-hole nozzle with an outlet diameter around 90 µm that has been extensively used for research [1–7], but its size has been specified to resemble characteristics of heavy-duty diesel injectors. Spray D have been specified with a convergent nozzle (*k-factor* = 1.5) with a broadly hydro-eroded inlet in order to suppress the cavitation phenomenon occurrence in its inner flow. Its nozzle shape have been measured through x-ray tomography and other methods by several authors such as Kastengren et al. [2] and is available online in CAD 3D-part format [8]. Particularly, the nozzle employed in this work, according to the ECN coding reference, is the #209103 that have an outlet diameter of 192 µm. This metric was measured at CMT and reported in [9] along with its mass and momentum flow rates. Figure 3.2 left picture shows the body of the injector, while image of the right is a tomography of the Spray D nozzle orifice obtained from [8]. The main features of the nozzle are summarized in Table 3.1

Fig. 3.2: ECN Spray D injector. Left: Bosch Spray D body. Right: X-ray tomographies of a Spray D nozzle [8].

Table 3.1: Characteristics for the Spray D nozzle [9].

Feature	**Spray D**	**Units**
Serial number	#209103	-
Reference name	D103	-
Nozzle outlet diameter	192	µm
k-factor	1.5	-
Steady-state mass flow$^{\triangledown}$	11.66	g/s
Steady-state momentum flux$^{\triangledown}$	6.81	N

$^{\triangledown}$ Value at $ET = 2.5\,\text{ms}$; $p_{rail} = 150\,\text{MPa}$; $p_{back} = 6\,\text{MPa}$

3.1.3 Electrical signal

For solenoid-driven injectors as the employed in this work, the needle is governed by three movement stages: the needle opening, the steady-period when the needle is completely opened and the closing. Regarding the electrical signal, an intensity peak is needed to start the movement of the needle, then an intermediate one is used to end this motion and a smaller intensity at the end to keep the needle lifted. As a large amount of variables are changed for parametric studies in this book, a large electrical signal of 2.5 ms was kept constant during all tests, and is shown in Figure 3.3. The specification ASOE in the horizontal axis stands for After Start of Energizing, which is the time reference from the triggering of the signal.

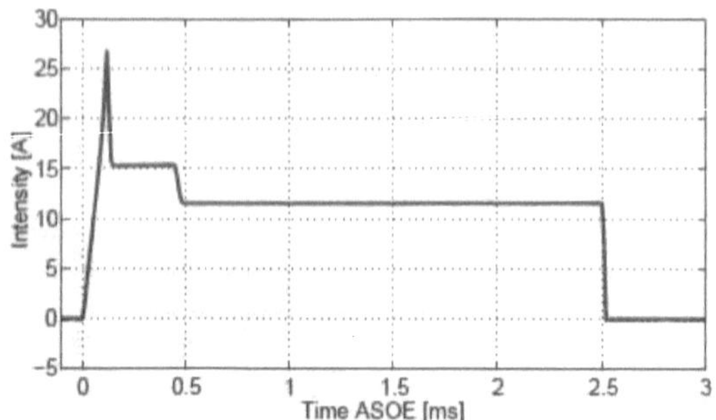

Fig. 3.3: Electrical signal used.

3.1.4 Fuels

n-dodecane and diesel #2 (name given in the industry accordingly to its refining grade) have been used as fuels in the different experiments that span this work. N-dodecane have been used in several past researches to represent diesel fuels in terms of mixing rate and ignition [10–14]. It was employed in this work since it is a mono-component fuel that has been widely characterized to be easily simulated in CFD, and it is also the primary fuel of study of the ECN working group. On the other hand, diesel #2 has been employed in order to have a more accurate representation compared with current commercial applications in engines and to study the effect of varying the fuel properties, which can be seen in Table 3.2.

3.2 High-temperature and high-pressure facility

All the experiments were performed in a high-pressure and high-temperature vessel (HPHTV from now). This test rig, according to the criteria introduced by Baert et al. [16], could be classified as a constant pressure and flow (CPF) test chamber. This optically-accessible is able to reach engine-like thermodynamic conditions with a continous gas flow at high-pressure and high-temperature, allowing to get nearly quiescent conditions, incrementing largely the maximum operating injection frequency that can be set, and reducing the time needed for the experiments. This facility has been employed previously in some researches [7, 14, 17–20]

Table 3.2: Fuel properties for n-dodecane and diesel #2 [8, 15].

Fuel Property	n-Dodecane	Diesel	Units
Reference name	nC12	D2	-
Total distillation temperature	489	623	K
Cetane number	87	46	-
Lower heat value	44.17	42.975	MJ/kg
Fuel density$^{\triangledown}$	752.1	843	kg/m^3
Aromatics concentration	0	27	%
H_2 mass concentration	15.3	13.28	%
Kinematic viscosity$^{\triangle}$	1.5	2.35	mm^2/s
Flash point	356	346	K
Sulfur content	0	9	ppm

$^{\triangledown}$ Value at 15 °C
$^{\triangle}$ Value at 40 °C

3.2.1 Description of the HPHTV

This facility is basically composed by four subsystems: gas compressors, heaters, test vessel and control panel. With different combinations of the valves shown in the test rig circuit in Figure 3.4, it can be operated in open and closed loop, depending on the oxygen concentration desired for the test, which is monitored using a lambda sensor. For standard air tests ($xO_2 = 0.21$, being xO_2 the oxygen concentration), the facility is run in open loop and the air is filtered, pressurized by the compressors and stored in the high-pressure tanks. Then, the air is delivered to a dryer, and enters to a 30 kW electric heater right in the entry of the vessel to control the ambient temperature inside. Hot air exits from the vessel and is cooled down before being released to the atmosphere. By operating in closed loop configuration, pure nitrogen or mixtures of O_2 and N_2 can be used as gases in order to recreate non-reactive conditions or to simulate exhaust gas recirculation (EGR) respectively. In order to do this, a nitrogen generator is used.

The chamber has a double layer configuration to improve the ambient temperature homogeneity, by avoiding big heat losses surrounding the internal thinner wall with an insulating material. This insulation covers the space to the external shell, which has the structural function of withstanding the ambient gas pressure. This configuration, provides the chamber with a test section of 200 mm. The facility has three optical accesses of 128 mm diameter placed orthogonally between them and is capable of reaching chamber conditions up to 1000 K and 13 MPa.

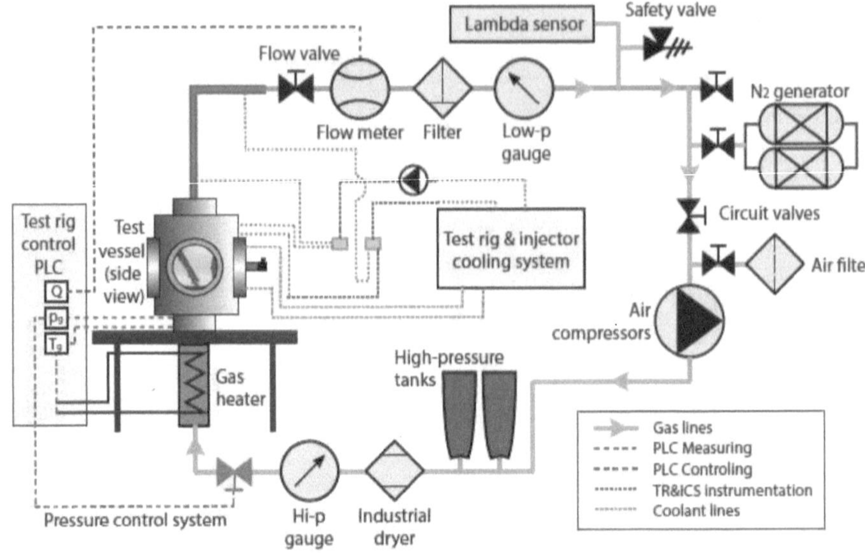

Fig. 3.4: Schematic diagram of the high pressure and temperature facility.

An electronic panel is used to set both in-chamber temperature and pressure. Both parameters are regulated by PID systems, that respectively control the gas heater and a pressure-regulation valve located upstream from the vessel and kept set-point values depending on the desired conditions. Furthermore, several safety parameters are monitored by this system, for example the gas flow (controlled manually by a valve downstream of the chamber), maximum heater output temperature, etc. The value of the gas temperature inside the chamber is estimated employing three thermocouples sheathed with the same material but with different diameters and extrapolating the gas temperature to a theoretical 0-mm-thermocouple, in order to suppress uncertainties produced by radiation or probe inertia. This approach is detailed in previous works [21, 22] and has been even used in other similar CPF rigs [23].

Some researches from CMT [20, 24] describe, particularly for this facility, the effect of fuel temperature on liquid length, and combustion parameters. A continuous flow of ethylene glycol is in direct contact with the injector (right part of the chamber in Figure 3.4) to keep under control its tip temperature and consequently the fuel temperature. A thermo-regulator unit provided with a PID system is the responsible of controlling the coolant temperature at a flow rate up to 0.30 l s $^{-1}$. This system feeds the liquid at a temperature range between 15 °C and 90 °C. The relation between the temperatures of the injector tip and the coolant has been determined in a previous work [20], where a description of the aluminum cap used in this work as a heat conductor between the coolant and the nozzle tip, is provided. Following the proceedings explained in that research, the fuel injection temperature was kept at 90 °C for all tests performed in this book.

3.3 Transparent wall system

The HPHTV, just as described in section 3.2, is perfectly employable for free spray characterization. Nevertheless, the design of additional hardware to adapt the facility to spray-wall interaction experiments was necessary. A transparent wall made of quartz (abbreviated in this work as QT-Wall) has been inserted in the interior of the chamber to visualize the spray impacting on it at temperature conditions near to the hot gases ones. To achieve that, the wall is significantly exposed to the hot gases of the vessel in all their faces. The entire supporting assembly has been attached to a new injector-holder cap, which has been designed with this purpose. The whole system has been carefully dimensioned to take advantage of the HPHTV inner volume and to be versatile enough to allow an easy change of wall position in terms of both distance from the injector tip and impingement angle. This supporting assembly has been described also in Payri et al. [19].

3.3.1 Supporting system description

Figure 3.5-left, shows the assembly used to hold the QT-Wall inside the vessel at the desired distance-angle configuration. A protective cap is employed as layer between the hot ambient and the injector body, and to support the rest of the wall system. Two folded sheets are screwed to the cap, and are used to hold two fixed 'U' shaped structures. Depending on this target position of the wall, a determined pair of exchangeable frames is used to support the wall holder. In Figure 3.5-right different geometries for those lateral frames are shown and how they are designed for different distances between the injector tip and the wall (d_w) and impingement angles (θ_w). Four hooks, tightened with screws, are used to press the JGS1 fused silica wall (100 mm × 60 mm × 10 mm) against a holder, which is screwed to the lateral frames. All pieces of this system, with clear exception of the wall, are made of stainless steel due to its ease for manufacturing and its durability and stability at high-temperatures.

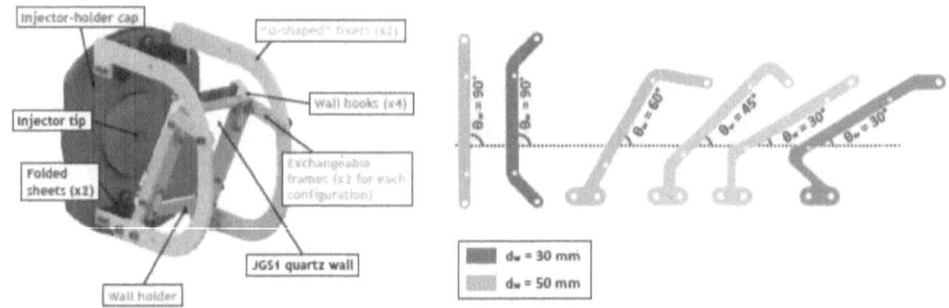

Fig. 3.5: Transparent wall support system. Left: Assembly with part labels. Right: Six lateral frame different geometries

The coordinate system used for the measurements is shown in Figure 3.6. The spray spreading is measured parallel respect to the wall surface while the spray thickness is measured in a normal-to-plate direction (y and z in Figure 3.6 left picture, where the positioning parameters can be seen in dashed red lines). Figure 3.6-right shows not only another view of the reference system (x axis and y proyected accordingly to the perspective with $sin(\theta_w)$) but also allows to appreciate the field of view through the transparent wall with a fixed width of 40 mm, in the case illustrated in the figure, for a $\theta_w = 60°$. The injector-holder cap has been painted in black matte for the experiments in order to avoid reflections of the flame light at combustion conditions.

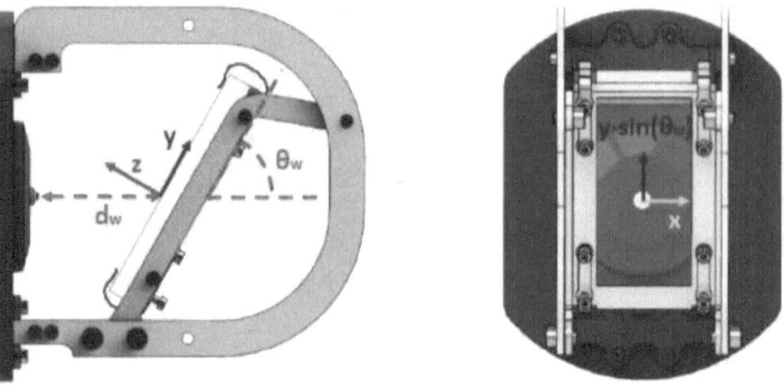

Fig. 3.6: Orthogonal views of the quartz wall assembly (d_w = 50 mm; θ_w = 60°). Left: Side view. Right: Front view.

3.4 Thermo-regulated and instrumented wall system

The aforementioned quartz wall is useful to focus the study on the macroscopic characteristics of the spray during the SWI through visualization, since it allows to optically access to the phenomenon through the wall, and to isolate it from the heat transfer between the spray and the wall surface. Nonetheless, the last is a process of significant interest taking into account the effect it has on the general performance of the engine [25–27]. In this innovative system (a thermo-regulated and instrumented wall or TRI-Wall), the possibility of a frontal view has been sacrificed in order to use a conductive metallic wall provided with fast-response probes to register the wall temperature instantaneously during the injection event. The wall was designed as a thermo-well in order to set its surface temperature by gas convection. A controlled flow of pressurized air previously was employed for that purpose on the cold face of the plate. In this case, the wall system is not attached to the injector holder, but it replaces the front window and, similarly as the quartz wall assembly does, it is capable of operating with different wall positions. The system could be divided into three sections: the air cooling circuit, the thermo-well, which is inserted into the chamber, and the temperature measuring system.

3.4.1 Air cooling section

The main intention of the thermo-well is to keep the wall at a controlled piston-like temperature, which could be in an approximated range between 430 and 600 K [28, 29], while the air of the surroundings is hotter. A coolant selection was made during the design phase of the TWI-Wall system, with two different main solutions in mind: air and liquid (water or oil). At the end, the options with liquid coolants were discarded fundamentally because of two reasons. First, liquid coolants provide high convective heat transfer coefficients which, accordingly to simulations made during this phase, could lead to lower wall temperatures than the requested or to have quite low flow requirements which could be difficult to manage. Additionally, the thermo-well design could be more complex just to prevent liquid leaks to the thermocouples wiring, taking into account the comprised dimensions of the system. On the other hand, air is a suitable alternative, considering the capacity of compressed air flow of the lab facilities and the low risk of damage to instrumentation.

Before introducing the air into the thermo-well to cool down the wall steadily, it is necessary to ensure it enters at adequate conditions of temperature and flow rate for this goal. Figure 3.7 shows a diagram of the air circuit used for that purpose. A compressor pressurizes atmospheric air and it is sent to a valve where the pressure can be regulated up to 6 bar. Then, the flow can be directed in three different forms depending on the selected combination of circuit valves: The air goes directly to the thermo-well, it goes entirely through a 50 L tank or it can be split in both paths in parallel regulating the flow by opening partially the valves. In this tank, solid CO_2 flakes are deposited which, since they sublimate at −78.5 °C (at 1 atm), are used to reduce air temperature. The flow is regulated with the valves in these possible configurations, not only to control accurately the wall temperature, but also to prevent the early evaporation of the dry ice. With this last purpose, the tank was wrapped uniformly with a thick layer of insulating material, minimizing heat transmission from the test room to the deposit.

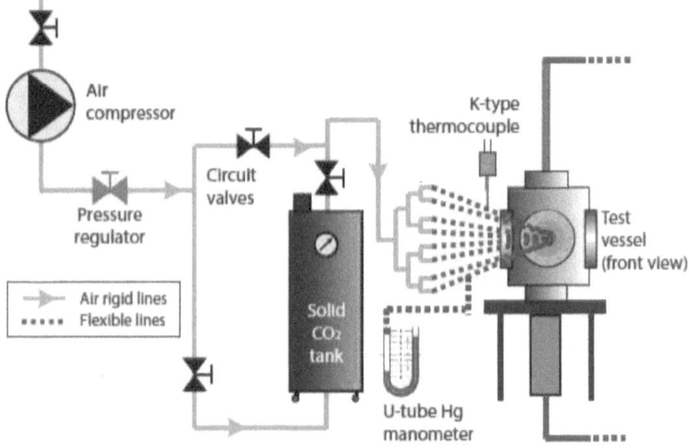

Fig. 3.7: Schematic diagram of the air cooling system for the thermo-well.

Downstream from the parallel pipes, the flow is divided into eight flexible lines with 6 mm internal diameter and 110 mm length, in order to distribute equally the flow in the thermo-well. The same kind of accessories and tubes have been used to ensure a better homogeneity in the flow of those latter lines. In order to know the air conditions right before its entrance in the thermo-well, a mercury-column manometer and a thermocouple are connected to two different lines from the eight ones. It is important to highlight that the dimensions of all the lines downstream from the convergence posterior to the dry ice tank, have been shortened in order to minimize losses of air pressure before the wall.

3.4.2 Thermo-well section

Figure 3.8 shows the entire thermo-well system, that has been designed to distribute the cold air and to cool down the wall. The wall temperature in the hot surface is intended to be homogeneous. An insulating 14 mm thickness layer is used to surround the thermo-well in order to prevent heat transfer from the cylinder side. It is important to highlight that, in contrast to the quartz wall system, this one has significant temperature and pressure differences at the two faces of the wall, therefore it had to be designed to stand those conditions and to be properly sealed. In Figure 3.8-right it is possible to observe the system in the inside. As seen, the thermo-well is a set of different pieces. This design offers several advantages such as an easier manufacture of pieces, more degrees of freedom for alignment and the capability of changing configurations. In the case of the wall, it is threaded to the rest of the thermo-well to be exchangeable and vary the impingement angle between two target conditions (60° and 90°).

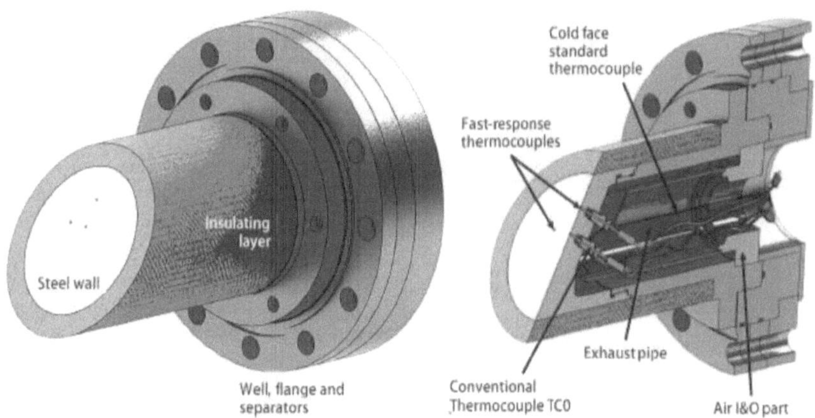

Fig. 3.8: Thermo-well assembly (d_w = 50 mm; θ_w = 60°). Left: Isometric view. Right: Isometric view with section cut.

The TRI-Wall is more detailed in Figure 3.9, along with the air flow. The wall is threaded to the well, union that is sealed with an aluminum gasket and a two-component silicone. The well is, as aforementioned, covered by an insulating layer that has been molded with that purpose and it is screwed to the well holder, that keeps the system in sealed contact with the window frame of the HPHTV. Nevertheless, the system is pressed against the test rig by 12 screws that are inserted in the flange and in the separator. In regards to the $\theta_w = 60°$ wall, respect to the perpendicular one that is shown in Figure 3.9, the wall is 15 mm axially longer for thermocouple insertion purposes (this can be noticed in Figure 3.10), making necessary the use of a second separator of this length to obtain the same d_w with both walls. This second separator, which is only used when the inclined wall is employed, can be seen in Figure 3.8. The separator shown in Figure 3.9 is the part that sets the wall distance from the injector tip, being located behind the flange for a closer wall configuration (d_w = 30 mm) or in front of it as shown in the picture for a d_w = 50 mm.

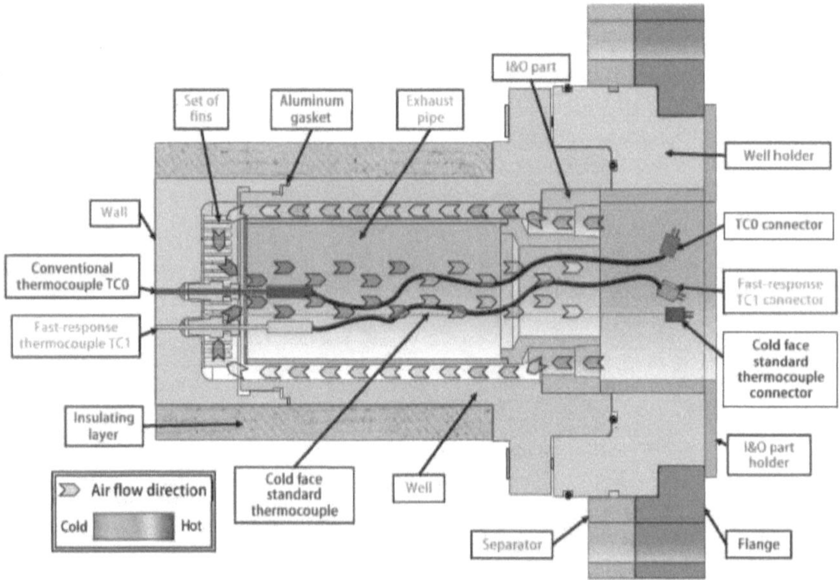

Fig. 3.9: Lateral view with transversal cut of the thermo-well (d_w = 50 mm; θ_w = 90°).

The flow path is governed by a piece created as air inlet and outlet (I&O part in Figure 3.8 and Figure 3.9) and an exhaust pipe. The I&O part has eight air inlets that are connected with brass fittings to the flexible tubes of cold air. This air enters in the thermo-well going through the external periphery of the exhaust pipe, until it reaches the wall. The pipe is provided with a sheet parallel to the wall which has been drilled in the center and in the location of the thermocouples, in order to prevent the immediate flow-back of the air. With this, the flow is in entire contact with the wall and extracts heat from it via convection, process that is enhanced with a set of aluminum fins. Afterwards, the hot air flows through the sheet holes and is released from the thermo-well across the inside of both the exhaust pipe and the I&O part. This piece is pushed against the well with a I&O part holder, screwed to the well holder. Even when the flow goes through the entire thermo-well, the insulating layer is used to ensure that the heat exchange is just significant in the wall, which together with the short time scale of the injection event, it allows the hypothesis of one-dimensional heat through the wall to calculate it from the temperature signal obtained from the fast-response thermocouples.

Those sensors, which are explained more carefully in the following section, are inserted in the wall through 1.6 mm diameter holes and attached to it with their corresponding NPT fittings (U.S. standard for conical threads known as National Pipe Thread), in order to reduce the back wall surface occupation. These thermocouples are located at different distances from the geometric center of the wall (d_{gc}), which is supposed to be defined as 'spray collision point'. Those distances are specified in Figure 3.10, where the two walls can be observed in both normal-to-wall and sectioned lateral views. Both the diameter of the perpendicular (circular) wall and the minor-axis of the inclined (elliptical) one are 94 mm and the wall thickness is 20 mm in both cases to ensure a similar heat conduction behavior. TC0 is always making reference to a standard K-type surface thermocouple with a conventional acquisition rate (non-fast) in order to have a robust reference sensor, while TC1 and TC2 are the names assigned to the two fast-response sensors. The conventional probe TC0 was selected to have the same dimensions and material than the fast-response ones.

Figure 3.11-left shows a real picture of the fins putted together onto the cold surface of the wall. They have been adhered to the wall using a highly conductive glue capable of standing temperatures up to 600 °C. The set is composed by a large amount of pieces as the illustrated in the Figure 3.11 right picture, which ensures all fins, even when they are from different pieces, are equally spaced. Several of them have been reshaped to follow the pattern of the wall and to leave room for the thermocouple fittings. Similarly, a groove has been made in three of them to place a bended conventional thermocouple between the fins and the wall; all this to have a temperature sensor in direct contact with the cold face of the wall and to monitor this boundary condition which is not susceptible to change as abruptly as the hot surface temperature with the spray-wall interaction, due to the thickness of the wall and the short event duration. All relevant dimensions of the fins are shown in Figure 3.11-right, where $\varnothing_f$ is the fin diameter, L_f is the fin length, d_f is the linear distance between neighbor fins and e_f is the thickness of the part base.

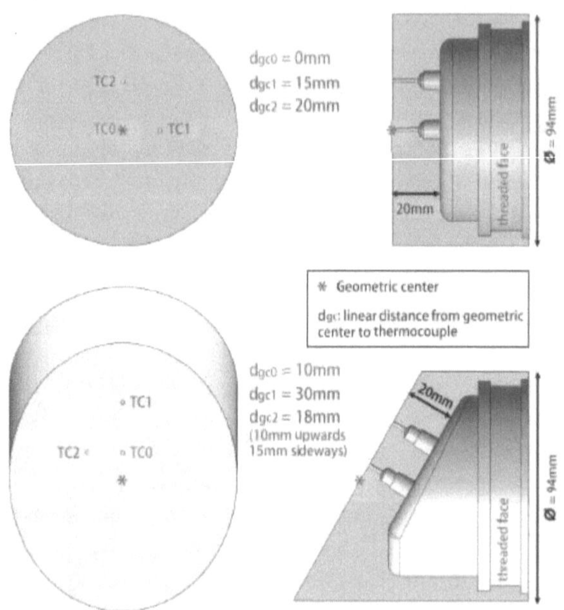

Fig. 3.10: Wall pieces in normal-to-wall view and lateral for both $\theta_w = 90°$ at the top and $\theta_w = 60°$ at the bottom (wall has been simplified being shown without fins nor threads)

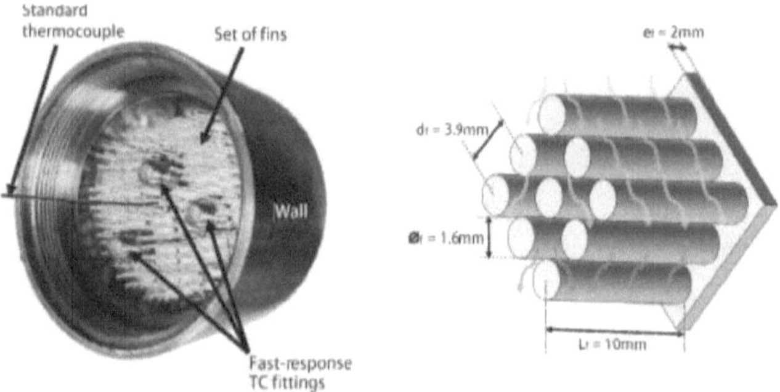

Fig. 3.11: Set of cooling fins. Left: Photo of fins mounted on the $\theta_w = 90°$ wall. Right: Detail of the fins.

3.4.3 Temperature measuring section

This last subsystem has the function of acquiring the information about the wall temperature evolution during the SWI event. Due to the reduced temporal scale of that event and the high resolution desired, the use of a conventional data acquisition system was discarded. Instead, the employed configuration can be seen in the scheme of Figure 3.12, where the connections are shown just for one thermocouple (TC1). The thermoelectric voltage measured is the produced by the difference between the hot junction of the circuit (the wall point where the probe is installed) and the cold junction (both copper unions with thermocouple materials, that are kept at 0 °C submerged into a ice bath, that acts simultaneously as isothermal block). The cold junction has been covered to avoid oxidation or wire damages. Then, both copper cables are connected to one high sensitivity channel of a Yokogawa DL750 ScopeCorder oscilloscope. This circuit, from the wall to the scope, is equivalent for all the three sensors.

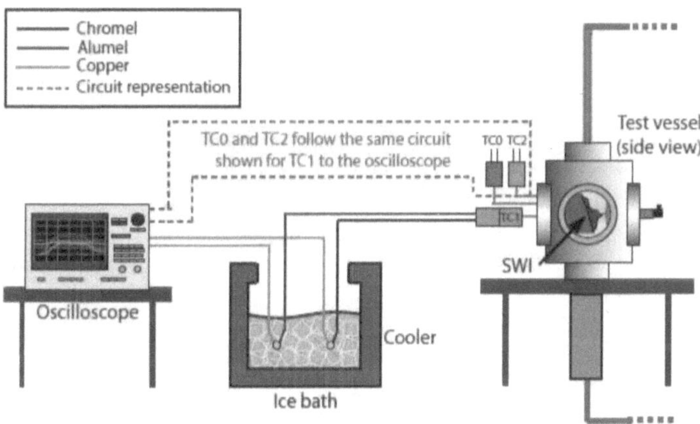

Fig. 3.12: Temperature measuring circuit. The connections are shown just for one of the three probes (one standard thermocouple and two fast-response thermocouples), being equivalent for the rest.

The most critical elements of this section are the two fast-response thermocouples used to acquire the temperature of the wall at a microsecond time-scale (TC1 and TC2). In the left image of Figure 3.13 one of the employed MedTherm TCS-061-K thermocouples is represented. Additionally, TC0 is a conventional thermocouple that is placed together with the fast ones, in order to have an additional check of the target surface temperature with a more robust sensor, always placed in the closest hole to the 'collision point', where the forces applied by the spray when it reaches the thermocouple, are the highest. Once the fitting is screwed in the wall, the probe is mounted flush in the wall surface. A lava gland is placed inside the fitting to tighten the probe in its axial position.

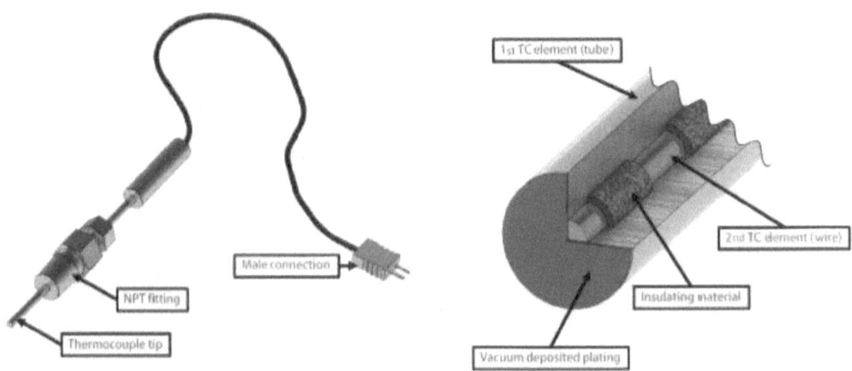

Fig. 3.13: Fast-response MedTherm TCS-061-K thermocouple (TC1 and TC2). Left: Image of the probe. Right: Detail of the junction.

The 1/16″ (1.6 mm) diameter fast thermocouples have a special junction design to reach high response and acquisition rates. This is detailed in a draw in Figure 3.13-right. This coaxial thermocouples consist of a small wire of one junction material, coated with a special ceramic insulation of 0.0127 mm thickness and high dielectric strength, which is swagged in a tube of the second junction material. The junction between both materials is made by vacuum depositing a metallic coating of 1-2 µm over the sensing extreme of the sensor, creating a metallurgical bond, as described by Lefeuvre et al. [30]. In this case a type-K thermocouple was selected due its reliability and its linearity range, therefore, the thermocouple is made of a chromel and alumel junction.

Before using the thermocouples mounted in the wall, a Fluke temperature bath system was used to calibrate them. The methodology was to expose the thermocouples to a dry bath into aluminum oxide sand fluidized by low pressure air. Bath temperature is governed by a PID-controlled heater. Since aluminum oxide is non-corrosive, non-flammable and has exceptional heat transfer properties, it allows to make safe and accurate calibrations, reaching stabilities of ±0.01 °C. Five calibration campaigns were made for the three probes in order to register the corresponding voltage profile for a temperature sweep and to check its linearity in a reasonable temperature range of testing. Figure 3.14 shows the profile of each thermocouple from averaging the raw signal of each calibration, along to the curve of a standard Type K thermocouple obtained from the NIST [31] (U.S. National Institute of Standards and Technology) which is identified as NIST-K. The image illustrates a very good agreement between the four curves, showing how the fast-response probes are calibrated respect to the conventional TC0 and the theoretical type-K curve. A high linearity of the voltage signal of the thermocouples respect to the temperature of the probes is observed too (R^2 = 0.9998 in the worst case), which is shown in the linear fit equations of the corner of the plot. This linear relationships were inverted to get the conversion from mV to °C for each probe. Direct voltage measurements were made during the tests using this conversion instead of a temperature measuring channel, in order to reach the highest possible acquisition rate, taking into account the short time scale of the diesel spray-wall interaction.

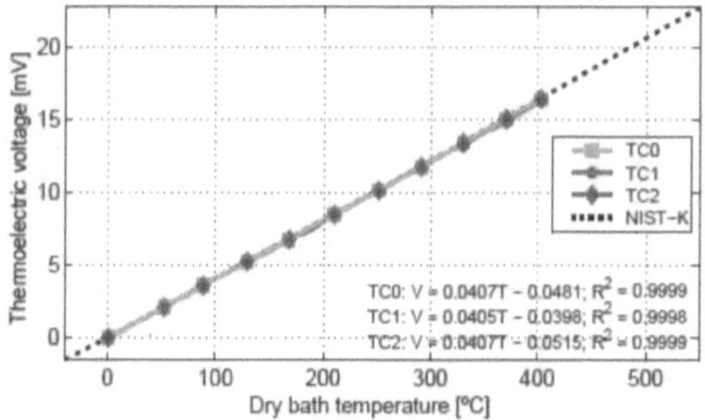

Fig. 3.14: Average calibration curves obtained for the three probes compared to the standard for type-K thermocouples. Equations obtained from linear fits of these data are shown in the right-bottom corner.

3.5 TRI-Wall thermal calculations

Prior the manufacturing of the thermo-well for the TRI-Wall, a preliminary simulation was made in order to verify the feasibility of the proposed system and to get an estimation of the temperatures when operating the HPHTV. To perform this study, the 'Flow Simulation' tool of SolidWorks® was employed. The domain was defined as the whole thermo-well from the cooling air inlet

(I&O part). Simulations were made at the most critical conditions of gas temperature and heat transfer for both assemblies: 90° and 60° walls.

3.5.1 Perpendicular wall at $\theta_w = 90°$

Figure 3.15-top shows a cut of the domain and the mesh levels for the fluid (left), solid (center) and all (right). A cubic-cell mesh has been defined with a maximum cell thickness of 8 mm and six different levels of refinement, reaching a minimum cell-size of 0.127 mm. The bottom part of the image shows a zoom where some of the most detailed cells are depicted. This refinement was initially made automatically by the software and afterwards, it was optimized by the user to obtain an adequate level of refinement in the wall, taken into account the behavior of the conduction phenomenon, the fins, and the air regions of high velocity and reduced flow section. Two lids, that are indicated in Figure 3.15, were added to define both the inlet and outlet boundary conditions of the air flow. The similar dimensions and material of all thermocouples fitted in the wall (one standard type-K and two fast-response ones), allowed to model them as equal, and the conventional thermocouple in the cold face of the wall was suppressed for not being considered to be significantly intrusive, similarly as the wires of TC0, TC1 and TC2. In total, a mesh with 6791645 elements has been used for this simulation.

The boundary conditions imposed were: an inlet mass flow of 43.2 g s^{-1} and temperature of 240 K, which were measured with the U-tube, the thermocouple and the pressurizing system of Figure 3.7 at 4 bar; an outlet pressure of 1 atm and a ambient (gas) temperature of 900 K which is the highest and nominal value in the test plan. A convective heat transfer coefficient of 300 W m^{-2} K^{-1} was assumed from previous thermal studies made in the HPHTV [32] and it was applied along to the mentioned T_{amb} in the in-vessel surfaces that are in contact with hot gases. The respective thermal properties of stainless steel, aluminum and the material of the isolation cover were applied. The simulation was set to run after reaching stabilization of bulk outlet mass flow, temperature and turbulent energy, bulk wall and fins averaged and max temperatures, and additionally after reaching 1335 iterations.

In Figure 3.16 the temperature plot of the wall surface can be seen. Even when a difference around 50 K is found between the hottest and the coldest cells, the temperatures measured at the thermocouple locations are within a range as wide as 12 K, which is considered acceptable for steady conditions. In detail, they are 528.07 K, 534.12 K and 540.51 K respectively. Furthermore, those temperatures are still close and below 550 K, which was the target wall

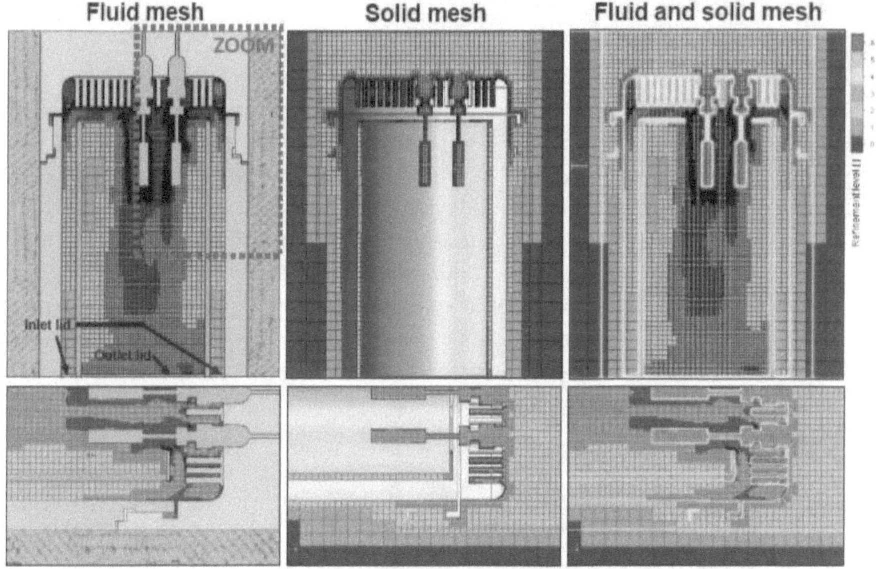

Fig. 3.15: Mesh of the computational domain used for the perpendicular wall simulation. Top: Entire domain. Bottom: Detailed view

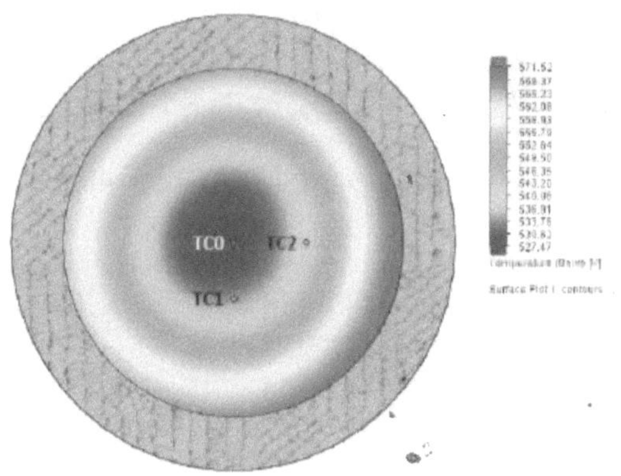

Fig. 3.16: Simulated profile of perpendicular wall surface temperature

temperature for the test points at $T_{amb} = 900\,\mathrm{K}$ and which is characteristic of diesel-like piston conditions [28, 29, 33, 34].

More interesting temperature information is depicted in the cut shown in Figure 3.17. The temperature profile of the center-plane across the vessel shows the heating of the air when goes through the fins and near to the wall to cool it down. The highest temperatures seem to appear in the insulating layer. In regards to the bulk wall temperature distribution, it can be seen how it is not perfectly linear. Nevertheless, it is barely non-unidimensional in the thermocouples region. Similarly, in the plot it is possible to observe some curves that represent the main streams of heat flux through the solid components. As expected, they are not completely horizontal, but they have an axial component strongly higher than the radial one. Furthermore, and more important for the assumption of 1-D heat, the streams are nearly parallel between them in the region near to the surface, which, taking into account how fast is the injection process, is the area of main interest.

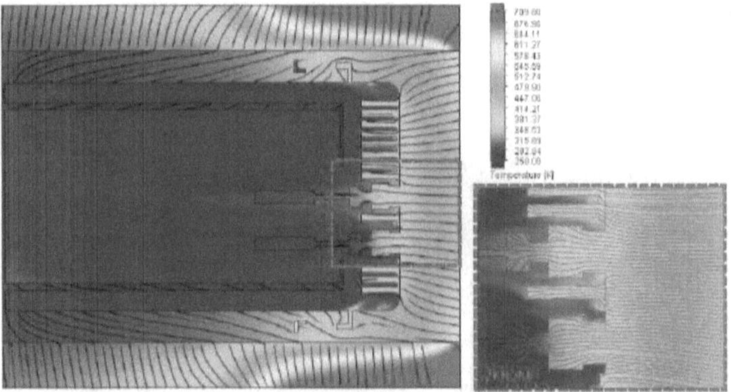

Fig. 3.17: Temperature distribution of a transversal cut of the perpendicular wall domain.

This information is quantitatively summed up in Table 3.3 for the first 1.5 mm depth inside the wall. As seen, the axial heat transfer is quite similar for all the thermocouples, being slightly lower as getting further from the center of the wall. Furthermore, as expected, the heat flux parallel to the wall is very low in the wall center and it gets higher at larger distances. The last two columns have a special interest for the validation of the hardware: even when the axial temperature gradients are important (around 15 K in 1.5 mm), the ratio between the axial heat flux in the surface ($\dot{q}_{z0}$) and the interior of the wall ($\dot{q}_z$) is near to 1 in all cases ($\dot{q}_z$ is always over 99% its value on the wall surface), which confirms the parallelism between the heat flux lines. For this case, which is the critical at the highest gas temperature, the heat component in the wall plane represents at most the 5.6% of axial component. That means that the 1-D heat calculated from the thermocouples temperature signal, is not disturbed by the depth of the wall and it is a well-representative simplification of the heat transfer for the designed system.

Table 3.3: Temperature and heat flux results from steady-state simulation of the perpendicular wall.

Probe	Depth through wall z	Temp. (solid)	Axial heat flux $\dot{q}_z$	Heat flux in wall plane $\dot{q}_{xy}$	$\dot{q}_{zo}/\dot{q}_z$	$\dot{q}_{xy}/\dot{q}_z$
Units	[mm]	[K]	[W m^{-2}]	[W m^{-2}]	[-]	[-]
TC0	0 (surface)	528.07	112785	2659	1	0.024
	0.5	522.39	113127	2686	0.997	0.024
	1	517.48	113461	2712	0.994	0.024
	1.5	513.22	113721	2702	0.993	0.025
TC1	0 (surface)	534.12	111200	5369	1	0.048
	0.5	528.73	111609	5400	0.996	0.048
	1	523.67	112022	5424	0.993	0.049
	1.5	519.56	112401	5468	0.991	0.049
TC2	0 (surface)	540.52	109167	6005	1	0.055
	0.5	535.13	109553	6068	0.996	0.055
	1	530.27	109935	6127	0.994	0.056
	1.5	526.81	110302	6133	0.991	0.056

3.5.2 Inclined wall at $\theta_w = 60°$

The same study was made for the inclined wall of 60°. A similar cell arrangement was used as depicted in Figure 3.18. In this occasion, 5491699 bricks are compounding the mesh in the whole domain, from sizes from 0.120 mm to 7.7 mm depending on the refinement level. A stabilization in the behavior of the converging parameters was obtained at 1355 iterations and the boundary conditions were replicated from the previous simulation of the perpendicular wall, due to both hardware are subjected to the same ambient and operating conditions.

Figure 3.19 shows the normal view of the wall with its temperature profile. As seen, temperature is considerably uniform except at the highest part of the wall. Nevertheless, besides the relatively small region of temperature gradients, the highest difference between thermocouples is approximately

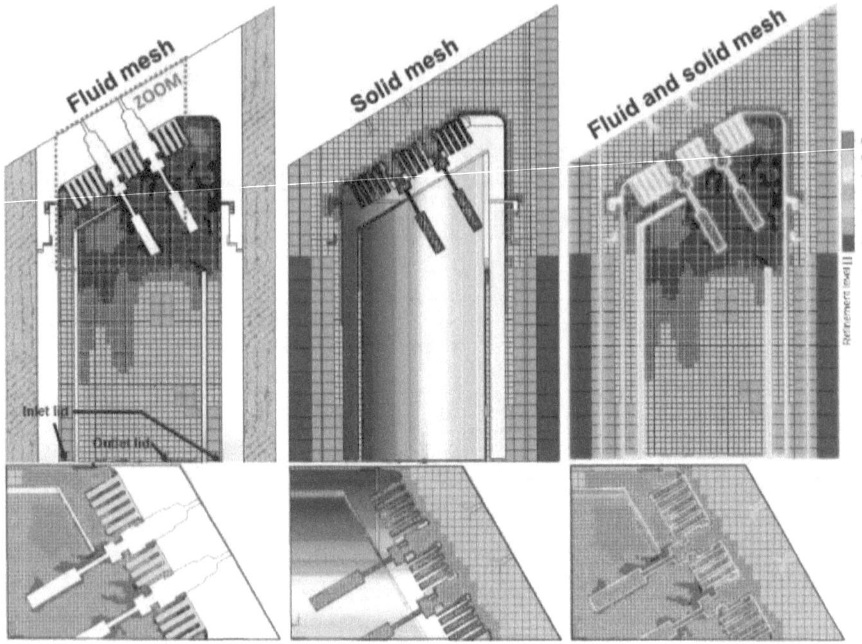

Fig. 3.18: Mesh of the computational domain used for the inclined wall simulation. Top: Entire domain. Bottom: Detailed view

20 K, nearly ±10 K around the setpoint surface temperature of 550 K. This demonstrates that the temperature uniformity of this wall is worse than the 90° case. However, the stable temperatures do not have to be the same at different points of the wall, taking into account that the coolant temperature is quite lower. This stronger temperature gradient between wall surfaces allows to pre-assume a heat flux with a predominant component oriented normal to the wall.

In Figure 3.20 the temperature map of the center-plane of the assembly is shown. Similarly as happens for the perpendicular wall, the highest temperatures have a trend to appear in the insulating cover. As noted in Figure 3.19, temperatures in both TC0 and TC1 locations are pretty well differenced. However, the lines that depict the heat flux main trajectories, are again, not normal to the wall but nearly parallel between them, with a stronger normal component. This can be even more noticed in the zoom of the figure.

Quantitative metrics of this are shown in Table 3.4. It is important to

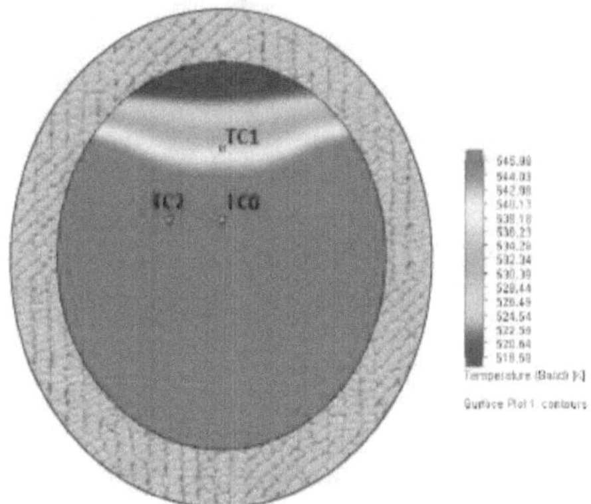

Fig. 3.19: Simulated profile of inclined wall surface temperature. The view is normal to the wall

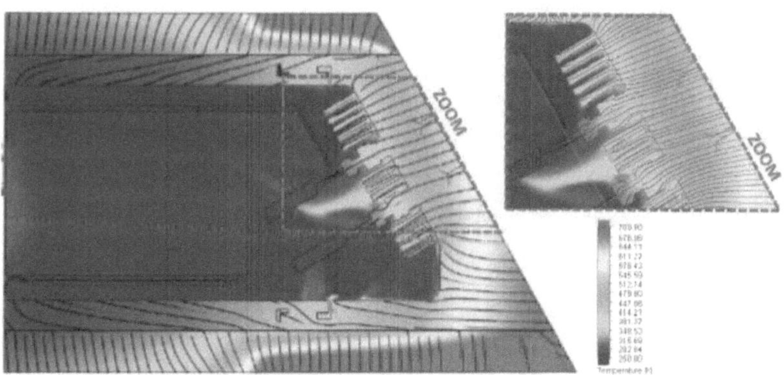

Fig. 3.20: Temperature distribution of a transversal cut of the inclined wall domain.

highlight that the values of the heat flux are projected orthogonally to the wall, being z the normal coordinate. As seen in this case, the highest variations of the normal feat flux $\dot{q}_z$ when go deeper into the wall up to 1.5 mm is still over the 97%, which indicates a good conservation in the steady heat transfer under a 1-D hypothesis. As it is shown later in chapter 4, just the first millimeters are relevant in the study of this heat transmission by SWI, due to the temporal scale of the event. Similarly, the relationship between the parallel-to-wall heat flux and the normal-to-wall one, is below 5% for the nearest thermocouple to the geometric center of the thermo-well. This difference rises up to 10.7% in the case of TC1, which could be expected as is in a zone of temperature parallel gradients (Figure 3.19). A maximum error of 9% is then made with the assumption of 100% 1-D heat transfer in TC1, however, the high $\dot{q}_{z0}/\dot{q}_z$ ratio indicates that variations or deviations of heat flux are practically negligible, ensuring that the signal obtained with TC1 has always a robust fundamental definition.

Table 3.4: Temperature and heat flux results from steady-state simulation of the inclined wall.

Probe	Depth through wall z	Temp. (solid)	Heat flux normal to wall $\dot{q}_z$	Heat flux in wall plane $\dot{q}_{xy}$	$\dot{q}_{z0}/\dot{q}_z$	$\dot{q}_{xy}/\dot{q}_z$
Units	[mm]	[K]	[W m^{-2}]	[W m^{-2}]	[-]	[-]
TC0	0 (surface)	555.59	93237	3488	1	0.037
	0.5	550.82	94148	2960	0.990	0.048
	1	545.71	94913	2707	0.982	0.049
	1.5	542.84	95844	2594	0.975	0.049
TC1	0 (surface)	537.59	135845	14564	1	0.093
	0.5	531.72	136412	13952	0.995	0.092
	1	529.27	137404	13552	0.988	0.088
	1.5	527.91	138001	13229	0.984	0.086
TC2	0 (surface)	562.63	101152	7449	1	0.073
	0.5	558.17	102248	7491	0.989	0.073
	1	554.32	103059	7490	0.981	0.072
	1.5	551.26	103894	7495	0.974	0.072

Even when both designs have been validated, a better understanding of their differences in wall heat transfer can be assessed by a fluid velocity study as the shown in Figure 3.21. SolidWorks® solves RANS (Reynolds Averaged Navier-Stokes) equations with k-ϵ turbulence model. As seen, the need of grooves in the exhaust pipe to have space for the thermocouples deviates in advance the air flow in the bottom part of the 60° wall, reducing the potential velocity of the air in this part near to the wall and affecting negatively the convection heat transfer in consequence. This deviation also does not allow to develop velocities between the sheet and the wall as high as they are for the perpendicular wall. A parallel-to-wall cut has been made at a half of the fins length, from their end. Not only velocities are dissimilar, as could be expected (higher for the 90° wall), but the velocity profile is more irregular for the inclined wall, as a result of the open holes in the sheet of the exhaust pipe. However, the flow of air do not have stagnation zones and it flows adequately between the fins to cool them down.

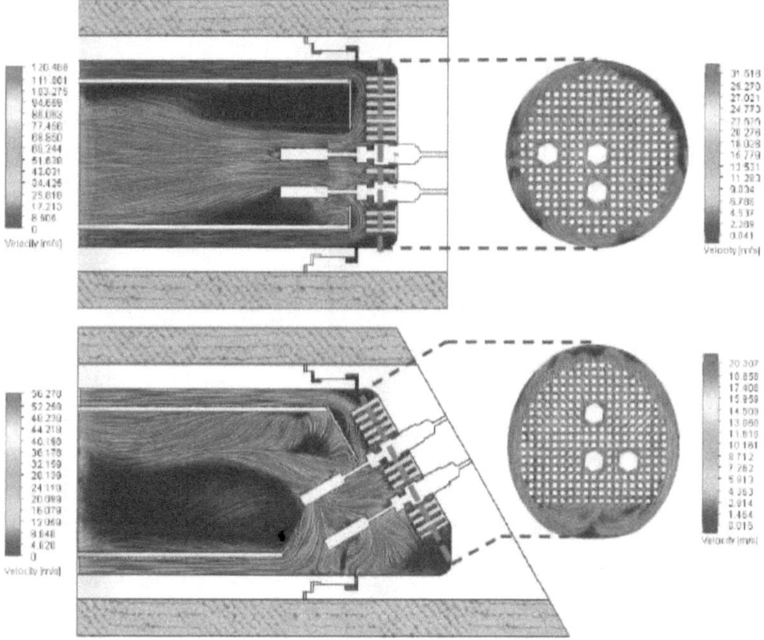

Fig. 3.21: Velocity profiles for both walls at their last iteration. Top: Perpendicular wall. Bottom: Inclined wall.

References

[1] Pickett, Lyle M, Genzale, Caroline L, Manin, Julien, Malbec, Louis-Marie, and Hermant, Laurent. "Measurement Uncertainty of

Liquid Penetration in Evaporating Diesel Sprays". In: *ILASS Americas 23rd Annual Conference on Liquid Atomization and Spray Systems.* Ventura, CA (USA): ILASS-Americas, 2011 (cited on page 46).

[2] Kastengren, Alan L et al. "Engine Combustion Network (ECN): Measurements of Nozzle Geometry and Hydraulic Behavior". In: *Atomization and Sprays* 22.12 (2012), pp. 1011–1052 (cited on page 46).

[3] Desantes, Jose Maria, Payri, Raul, Gimeno, Jaime, and Marti-Aldaravi, Pedro. "Simulation of the First Millimeters of the Diesel Spray by an Eulerian Spray Atomization Model Applied on ECN Spray A Injector". In: *SAE Technical Paper 2014-01-1418.* 2014 (cited on page 46).

[4] Jung, Yongjin, Manin, Julien, Skeen, Scott A, and Pickett, Lyle M. "Measurement of Liquid and Vapor Penetration of Diesel Sprays with a Variation in Spreading Angle". In: *SAE Technical Paper 2015-01-0946* (2015) (cited on page 46).

[5] Xue, Qingluan et al. "An Eulerian CFD model and X-ray radiography for coupled nozzle flow and spray in internal combustion engines". In: *International Journal of Multiphase Flow* 70 (2015), pp. 77–88 (cited on page 46).

[6] Meijer, Maarten et al. "Engine Combustion Network (ECN): Characterization and comparison of boundary conditions for different combustion vessels". In: *Atomization and Sprays* 22.9 (2012), pp. 777–806 (cited on page 46).

[7] Benajes, Jesus, Payri, Raul, Bardi, Michele, and Martí-aldaraví, Pedro. "Experimental characterization of diesel ignition and lift-off length using a single-hole ECN injector". In: *Applied Thermal Engineering* 58.1-2 (2013), pp. 554–563 (cited on pages 46, 48).

[8] ECN. *Engine Combustion Network.* Online. 2010 (cited on pages 46, 47, 49).

[9] Payri, Raul, Gimeno, Jaime, Cuisano, Julio, and Arco, Javier. "Hydraulic characterization of diesel engine single-hole injectors". In: *Fuel* 180 (2016), pp. 357–366 (cited on page 47).

[10] Linne, Mark. "Imaging in the optically dense regions of a spray: A review of developing techniques". In: *Progress in Energy and Combustion Science* 39.5 (2013), pp. 403–440 (cited on page 48).

[11] Falgout, Zachary, Rahm, Mattias, Sedarsky, David, and Linne, Mark. "Gas/fuel jet interfaces under high pressures and temperatures". In: *Fuel* 168 (2016), pp. 14–21 (cited on page 48).

[12] Skeen, Scott A et al. "A Progress Review on Soot Experiments and Modeling in the Engine Combustion Network (ECN)". In: *SAE International Journal of Engines* 9.2 (2016) (cited on page 48).

[13] Payri, Raul, Viera, Juan Pablo, Gopalakrishnan, Venkatesh, and Szymkowicz, Patrick G. "The effect of nozzle geometry over internal flow and spray formation for three different fuels". In: *Fuel* 183 (2016), pp. 20–33 (cited on page 48).

[14] Payri, Raul, Viera, Juan Pablo, Gopalakrishnan, Venkatesh, and Szymkowicz, Patrick G. "The effect of nozzle geometry over ignition delay and flame lift-off of reacting direct-injection sprays for three different fuels". In: *Fuel* 199 (2017), pp. 76–90 (cited on page 48).

[15] Kook, Sanghoon and Pickett, Lyle M. "Liquid length and vapor penetration of conventional , Fischer-Tropsch , coal-derived , and surrogate fuel sprays at high-temperature and high-pressure ambient conditions". In: *Fuel* 93 (2012), pp. 539–548 (cited on page 49).

[16] Baert, Rik S G et al. "Design and operation of a high pressure, high temperature cell for HD diesel spray diagnostics: guidelines and results". In: *SAE paper 2009-01-0649* 4970 (2009) (cited on page 48).

[17] Payri, Raul, Gimeno, Jaime, Bracho, Gabriela, and Vaquerizo, Daniel. "Study of liquid and vapor phase behavior on Diesel sprays for heavy duty engine nozzles". In: *Applied Thermal Engineering* 107 (2016), pp. 365–378 (cited on page 48).

[18] Payri, Raul, Salvador, Francisco Javier, Manin, Julien, and Viera, Alberto. "Diesel ignition delay and lift-off length through different methodologies using a multi-hole injector". In: *Applied Energy* 162 (2016), pp. 541–550 (cited on page 48).

[19] Payri, Raul, Gimeno, Jaime, Peraza, Jesús E., and Bazyn, Tim. "Spray / wall interaction analysis on an ECN single-hole injector at diesel-like conditions through Schlieren visualization". In: *Proc. 28th ILASS-Europe, Valencia* September (2017) (cited on pages 48, 51).

[20] Gimeno, Jaime, Martí-Aldaraví, Pedro, Carreres, Marcos, and Peraza, Jesús E. "Effect of the nozzle holder on injected fuel temperature for experimental test rigs and its influence on diesel sprays". In: *International Journal of Engine Research* 19.3 (2018), pp. 374–389 (cited on pages 48, 50, 51).

[21] Pitts, William M. et al. "Temperature Uncertainies for Bare-Bead and Aspirated Thermocouple Measurements in Fire Environments". In: *Thermal Measurements: The Foundation of Fire Standards, ASTM STP 1427. American Society for Testing and Materials (ASTM).* 2001 (cited on page 50).

[22] Bardi, Michele. "Partial needle lift and injection rate shape effect on the formation and combustion of the Diesel spray". PhD book. Valencia (Spain): Universitat Politècnica de València, 2014 (cited on page 50).

[23] Fitzgerald, Russell P, Svensson, Kenth, Martin, Glen, Qi, Yongli, and Koci, Chad. "Early Investigation of Ducted Fuel Injection for Reducing Soot in Mixing-Controlled Diesel Flames". In: *SAE Technical Paper 2018-01-0238* (2018), pp. 1–17 (cited on page 50).

[24] Payri, Raul, Garcia-Oliver, Jose Maria, Bardi, Michele, and Manin, Julien. "Fuel temperature influence on diesel sprays in inert and reacting conditions". In: *Applied Thermal Engineering* 35 (2012), pp. 185–195 (cited on page 50).

[25] González, Uriel. "Efecto del choque de pared en las características del chorro Diesel de inyección directa". PhD book. Valencia: E.T.S. Ingenieros Industriales. Universitat Politècnica de València, 1998 (cited on page 53).

[26] Meingast, Ulrich, Staudt, Michael, Reichelt, Lars, and Renz, Ulrich. "Analysis of Spray / Wall Interaction Under Diesel Engine Conditions". In: *SAE Technical Paper 2000-01-0272* 724 (2000), pp. 1–15 (cited on page 53).

[27] Köpple, Fabian et al. "Experimental Investigation of Fuel Impingement and Spray-Cooling on the Piston of a GDI Engine via Instantaneous Surface Temperature Measurements". In: *SAE International Journal of Engines* 7.3 (2014), pp. 2014–01–1447 (cited on page 53).

[28] Wang, Peng et al. "The flow and heat transfer characteristics of engine oil inside the piston cooling gallery". In: *Applied Thermal Engineering* 115 (2017), pp. 620–629 (cited on pages 53, 64).

[29] Mayer, Daniel et al. "Experimental Investigation of Flame-Wall-Impingement and Near-Wall Combustion on the Piston Temperature of a Diesel Engine Using Instantaneous Surface Temperature Measurements". In: *SAE International Journal of Engines.* Vol. 7. 3. 2018, pp. 2014–01–1447 (cited on pages 53, 64).

[30] Lefeuvre, T., Myers, P. S., and Uyehara, O. A. "Experimental Instantaneous Heat Fluxes in a Diesel Engine and Their Correlation". In: (1969) (cited on page 60).

[31] NIST. *National Institute of Standards and Technology standard reference data.* Online (cited on page 61).

[32] Ganeau, Louise. "Internal nozzle flow simulation of a high pressure and high temperature vessel". Internship Report. INSA Rouen Normandie, 2017 (cited on page 62).

[33] Kajiwara, Hidehiko, Fujioka, Yukihiro, Suzuki, Tatsuya, and Negishi, Hideo. "An analytical approach for prediction of piston temperature distribution in diesel engines". In: *JSAE Review* 23.4 (2002), pp. 429–434 (cited on page 64).

[34] Zhang, Hongyuan, Xing, Jian, and Guo, Chang. "Thermal analysis of diesel engine piston". In: *Journal of Chemical and Pharmaceutical Research* 5.9 (2013), pp. 388–393 (cited on page 64).

Chapter 4

Experimental and data processing methodologies

While the previous chapter describes most of the hardware employed to carry out the data acquisition; the methodologies for both, laboratory work and data processing are explained in this part of the book. The first section of this chapter is driven to explain the optical techniques employed in the experiments to record different kinds of spray information. The following two sections are related to the processing of the data obtained from the lab campaigns: the first of them details the image processing routines used to extract metrics from the images of the cameras used in the previous section, and the last one is related to the treatment of the signal obtained from the fast-response thermocouples to get the surface temperature profile and to compute the heat flux in the wall.

4.1 Optical techniques

Several diagnostics can be made via optical techniques depending on the selected set of equipment. Despite the fact that making a phenomenon optically accessible often needs the use of purpose-build facilities and makes the conditions further from the existing into commercial hardware, optical techniques are a strong and non-intrusive approach to get relevant information, even in many fields. In diesel spray research, the vapor macroscopic characteristics of a spray with high temporal resolution [1, 2], liquid length

[3–5], droplet diameter [6, 7] and velocity [8], projected fuel mass [9, 10], soot concentration [11–13], equivalent ratio [14], temperature distribution [15], ignition delay [16, 17] and lift-off length [18–20] are several examples of variables of interest obtained from optical techniques. Even internal flow characteristics such as cavitation regime [21] or needle-lift [22, 23] can be optically accessed.

In this work, four different optical techniques were employed for spray visualization. Single-pass Schlieren imaging was used to get the contour and macroscopic characteristics of the vapor phase of the spray in both inert and reactive conditions, as well as ignition delay when combustion is present. At non-reactive conditions, the liquid length was determined using a Diffused Back-Illumination (DBI) setup. The flame evolution was recorded from different angles by filtering the natural luminosity of the flame and, lastly, an intensified camera was added to the two setups for reacting sprays, in order to measure lift-off length. This section explains the principal concepts behind the techniques and then, it details the optical arrangements for the different campaigns that were used. These setups were designed in order to use two or more cameras at the same time in order to record different kind of information simultaneously from the same event, improving the consistency of the data and making a significant reduction of the time required to perform the experiments.

4.1.1 Single-pass Schlieren imaging

Vapor macroscopic characteristics of sprays have been extensively studied by engine researchers through Schlieren imaging [24–29]. This technique, which is also extensively used in other applications such as photographing of atmospheric disturbances produced by high speed projectiles, sound speed estimations or detection of heated gases or pressure waves [30], is based on the relationship between density and refractive index [31]. When light beams go through any path, they are deflected proportionally to the refractive index of the medium n_r. Equation 4.1 shows the behavior of this deflection ε of a light ray that travels through an optical path with L length in a generic axis u through a medium with a refractive index n_{r0}. This expression takes into account the density gradients in the surroundings.

$$\varepsilon_u = \frac{L}{n_{r0}} \cdot \frac{\partial n_r}{\partial u} \tag{4.1}$$

In research of sprays, this technique is applied commonly to visualize the boundaries of the spray vapor phase due to the deflection of the light beams produced by density gradients between the injected fuel and the ambient gas, in accordance with Equation 4.1 [32]. With this purpose, the two most widely employed configurations are single-pass and double-pass Schlieren. While the last of them is mainly used for multi-hole injectors or facilities with just one optical access available, the single-pass setup is commonly employed for single-hole axially drilled nozzles. This, and the low interference of this configuration with the simultaneous use of another optical techniques, made single-pass Schlieren the selected configuration for vapor spray visualization in this work. Figure 4.1 illustrates the principle of a standard single-pass Schlieren configuration. The image is divided into both top and bottom parts to represent different kinds of light sources. Beginning with the single point source case (bottom) we can observe how the rays are collimated by a lens. The non-deflected beams, represented in the figure by thin dotted lines, converge after the second lens in its focal distance, being able to pass through the slit and reach the screen where the observer (a camera or sensor) is located. On the other hand, the rays affected by the density gradients in the test zone (red dashed lines), are deflected an angle ε and are blocked by the slit. While this theoretically would produce a black and white image, the real light source is not an infinitesimal point but a finite source. This case, illustrated in Equation 4.1 top half, can be considered as a set of point sources, which generates a gray-scaled composition of images on the screen.

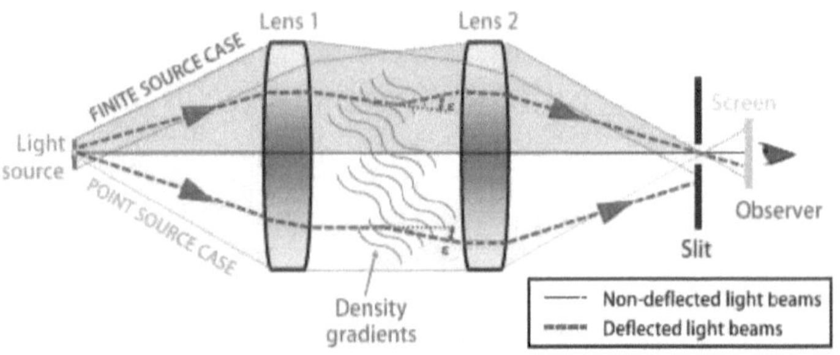

Fig. 4.1: Scheme of the Schlieren technique principle. Bottom half in yellow represents the point light source while top half in green shows the finite source case.

4.1.2 Diffused back-illumination

Diffused back-illumination or DBI is frequently used in spray diagnostics to get information about the air-fuel mixing process through determining the liquid penetration of the jet [33]. In this method, as its name indicates, the background is illuminated by a diffused light, which is extincted or blocked by the spray due to its optical thickness τ. The relationship between this number and the image intensity is given by the Beer-Lambert law in the following form:

$$\frac{I}{I_0} = e^{-\tau} \tag{4.2}$$

Where I is the intensity distribution of the image a of determined instant that considers extinction produced by the spray, and I_0 is the intensity of a reference image from a diffused source without any blocking or attenuation. Similarly, another definition of τ is through the length crossed by the light through the spray L_s and the extinction coefficient K_{ext} which depends on the spray or the attenuating body composition:

$$\tau = K_{ext} \cdot L_s \tag{4.3}$$

Many works in literature [1, 3, 24, 34, 35] employ fast light-emitting diodes (LED) technology to enhance light throughput, by using a short pulse instead a continuous light source. This approach, used in the experiments of this manuscript, allows to manage the light that enters in the camera sensor by setting the light pulse to very short durations.

The diffused illuminated background can be achieved by several optical setups. The principle used in this work is based on the concept proposed by Ghandhi and Heim [36] and is shown in Figure 4.2, where an engineered diffuser creates a radiant intensity distribution from a collimated input. A Fresnel lens is used then to collect the light and send it to the observer. This type of setup provides spatial uniformity of the intensity distribution with the drawback of the required diffuser size and arrangement dimensions.

4.1.3 Broadband flame chemiluminescence

Chemiluminescence is defined as the emission of photons due to chemical reactions. Combustion produces high-energy species whose excited energy level decays afterwards, releasing light when the final products are obtained [37, 38]. Chemiluminescence strongly occurs in different wavelengths depending on the products or radicals that are decaying to a low-energy state [37, 39] and therefore, soot incandescence rules over the visible part of the spectrum of luminescence produced in diffusive flames. Accordingly to this, Figure 4.3 shows the emissivity for different reaction species. Several researchers have not only employed flame chemiluminescence in a wide band

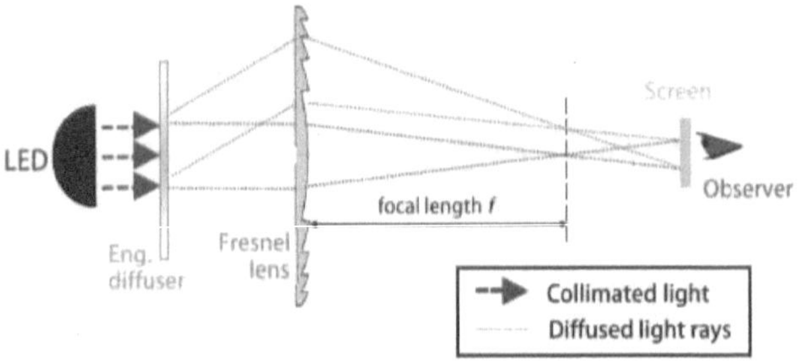

Fig. 4.2: Schematics of the DBI setup presented in [36].

to visualize its natural luminosity and the flame macroscopic shape, but it has been used also to determine both the location of the combustion start and ignition delay [16, 25, 40].

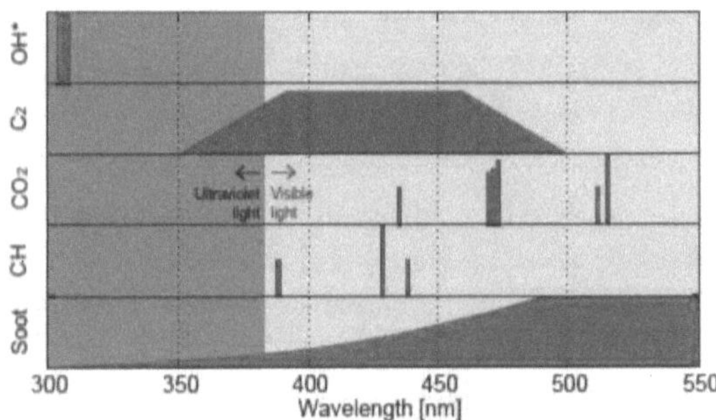

Fig. 4.3: Relative flame emission wavelengths for some common combustion radicals. Based on the presented in [39].

4.1.4 OH* chemiluminescence

As mentioned, soot luminosity is predominant in direct flame visualization on a wide range of visible wavelengths respect to other combustion products. However, its formation not only takes time but also its concentration is not uniformly representative for diesel flames stabilization. According to Peters [41] OH* radicals chemiluminescence takes place at the stable flame length, under high-temperature conditions, making them more suitable to measure the lift-off length (*LoL*) [42]. As seen in Figure 4.3, the light emitted by OH* radicals is inside the range of ultraviolet rays (UV rays) and embraces a narrow band of high emittance between 306-315 nm. This range, distant from most wavelengths of other flame radicals, is ideal to use compressed-range filters to block interference of other species, as it has been implemented by several authors [17, 42–45].

4.1.5 Optical setup: Quartz wall

From the understanding of the physical principles of the optical techniques, the optical setups were arranged. A relevant criterion was to use simultaneously the higher amount of techniques to suppress the necessity of repeating campaigns and to ensure all cameras are recording the same event at the same conditions. Additionally, interference between techniques or big influence of the optics on the assemblies within the vessel were avoided.

Having said that, the optical setup design was ruled by the wall hardware and the presence of combustion in the experiment. In the case of the quartz wall, the same two arrangements used for inert and reactive conditions respectively, were employed in the preliminary free-jet test performed as reference. Figure 4.4 shows the optical arrangement used for this wall at non-reactive condition (full nitrogen atmosphere). The parallel rays for Schlieren are achieved using a white-light 150 W halogen lamp with an incorporated cut-off to make a 'point' light source, and a collimating mirror. The incidence angle between the light source and the mirror was minimized in order to reduce beam straightening. After that, the parallel beams go through

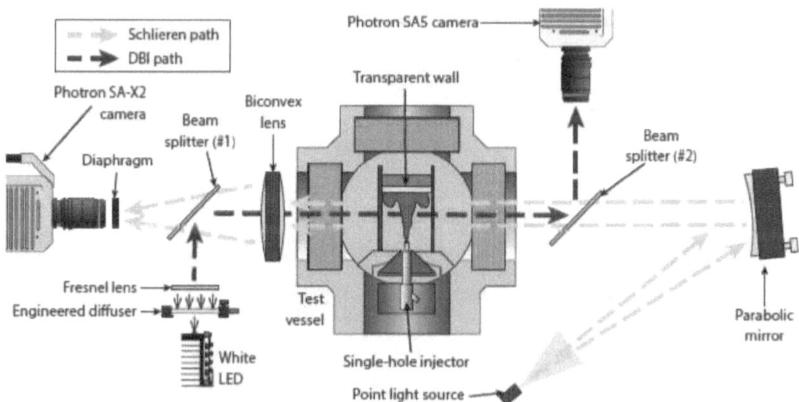

Fig. 4.4: Optical setup employed for both liquid and vapor phases visualization using the QT-Wall and also at free-jet conditions. DBI light main path is indicated in blue dashed arrows, while the yellow ones are illustrating the Schlieren beams path

the testing region, deviating the ones affected by the density gradients in the chamber. The rays are collected by a 150 mm lens with 450 mm focal length, distance where a diaphragm is placed just before a Photron SA-X2 high-speed camera, to act as slit to block the deflected rays. This path followed by the Schlieren technique is shown in Figure 4.4 in yellow dashed arrows, and it is similar to the implemented in the whole work for spray vapor phase visualization.

In regards to liquid length, the DBI rays main trajectory is depicted in Figure 4.4 in blue dashed arrows. The white LED rays are sent to an engineered diffuser which homogenizes the background and then to a fresnel lens which magnify the light intensity. Then, there are two 50/50 beam splitters (one before and a second one after the test region) to avoid interference with Schlieren imaging, which direct the beams towards the chamber and finally to the Photron SA5 camera in which the liquid phase videos are recorded. As mentioned in the DBI section, the LED pulse duration was minimized, in this case to 57 ns in order to improve the image quality. Both cameras are set to start acquisition with injector energizing.

This last DBI imaging was suppressed for reactive spray conditions, and was replaced with three other optical approaches to detect relevant combustion characteristics, as visible in the setup of Figure 4.5. Schlieren technique is used similarly as aforementioned to observe the gases of the reactive spray, with the addition of a 480 CWL (center wavelength) band-pass filter to prevent image

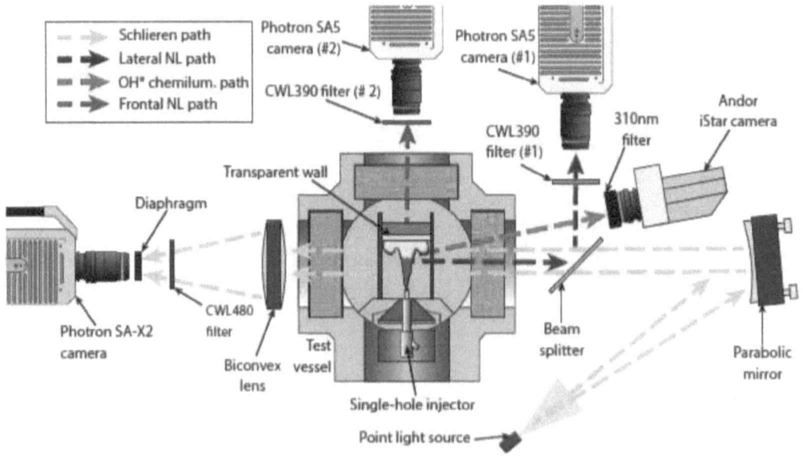

Fig. 4.5: Optical setup employed when using the quartz wall in a reacting atmosphere. The main trajectories by technique are shown in accordance with the legend.

saturation due to soot incandescence. Direct flame luminosity visualization from a side view is possible by the use of a beam splitter, as is shown in Figure 4.5 in blue arrows, using a 390 CWL filter. In the same way, natural luminosity is recorded with a high-speed camera, from the frontal optical access through the transparent wall, as is shown with green dashed arrows. To visualize the OH* chemiluminescence, an intensified charge-coupled device (ICCD) camera provided with a 100 mm f/2.8 UV-capable lens was employed. A 310 ± 5 nm band-pass filter is used to block different wavelengths to the OH* emissivity one. This light trajectory is shown in dashed red lines in Figure 4.5. The optical setup for the free-jet tests was the same, just omitting the use of a frontal camera for natural luminosity (NL) recording.

4.1.6 Optical setup: Thermo-regulated wall

The TRI-Wall was just used for reactive tests. Due to the different hardware inside the vessel, the optics had to be adapted for that changes. Figure 4.6 shows the optical arrangement employed were three cameras are used simultaneously to record the injection and the SWI events. The lateral flame visualization is made in the same way shown in Figure 4.5. Nevertheless, to ensure the perpendicularity in lateral visualization of lift-off length, in this opportunity it is made from the same optical access of the Schlieren camera. Therefore, the regular biconvex lens had to be changed by an UV lens, capable

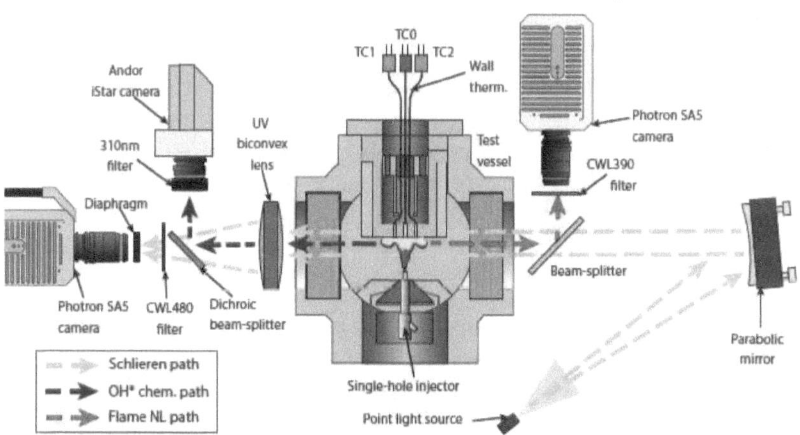

Fig. 4.6: Optical setup used in the tests with the TRI-Wall. Schlieren rays path is in yellow dashed arrows. Blue arrows show the main path of the OH* UV rays to the ICCD camera. Natural luminosity of the flame follows the way indicated in red arrows to its respective Photron camera.

of collecting ultraviolet light. After that, a dichroic beam splitter is used to let the visible rays pass through it to the Schlieren high-speed camera while the UV rays are reflected to the intensified camera (Andor iStar).

4.2 Chamber temperature characterization

Before performing the tests, two concerns respect to the homogeneity of the in-chamber temperature appeared: strong gradients on ambient temperature would not allow to have an effective control on that parameter, incrementing experimental uncertainty and having undesired effects on spray evaporation and ignition and, as seen in the previous section, the Schlieren technique is based on the refraction of light due to density gradients, which could be associated to temperature irregularities. Strong temperature inhomogeneities inside the vessel could affect the suitability of the use of Schlieren for spray visualization. This study was made in [17] using the HPHTV for a free-jet campaign (test rig without wall and three optical accesses) obtaining an homogeneous temperature profile. Nevertheless, the insertion of both walls required the verification of this condition. This was made for both wall hardware at $\theta_w = 90°$ and two tip-wall distances ($d_w = 30\,\mathrm{mm}$ and $d_w = 50\,\mathrm{mm}$).

Figure 4.7 depicts the setup employed to characterize the temperature along the spray axis. Accordingly to the left picture, one side window was replaced by a metallic 'dummy' window, which is provided with a set of thermocouples installed flush to the plume axis and at the same vertical level. Thermocouples where located as shown in Figure 4.7-right, in different positions respect to the distance between the wall and the injector tip. Three probes (pink crosses) were used in all characterizations. In the case of the TRI-Wall, wall surface temperature was measured with the nearest mounted probe respect to the axis. As the QT-Wall is not provided with a thermocouple, an additional one was put on the surface of the wall (shown as a blue circle mark). No injections were made in this campaign.

Results of this study are summarized in Figure 4.8. In the case of the QT-Wall (left plot), different ambient temperatures were set, while for the TRI-Wall it was possible to reach different combinations of T_{amb} and T_w to evaluate the effect of wall cooling. As seen, the temperatures are quite stable along the injector axis and the drop of the temperature in the quartz wall surface is never over 6 K, which is considered adequate taking into account that the minimum gas temperature in all test plans is 700 K. In the case of the cooled wall, the temperature is still 3 K around the setpoint T_{amb} for

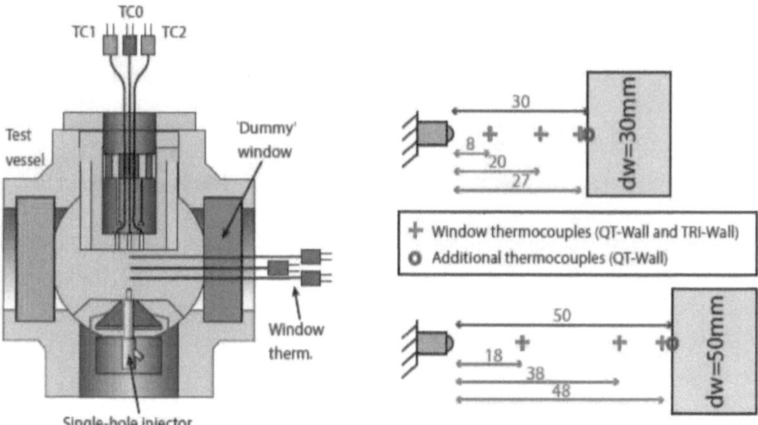

Fig. 4.7: Setup to measure the on-axis temperature profile in the vessel. Left: View of the vessel (scheme made for the TRI-Wall). Right: Location of the window thermocouples (please note these distances are in millimeters).

the window thermocouple at few millimeters from the wall (2 mm or 3 mm depending on d_w). Lastly, temperature measured in the wall is quite low as set.

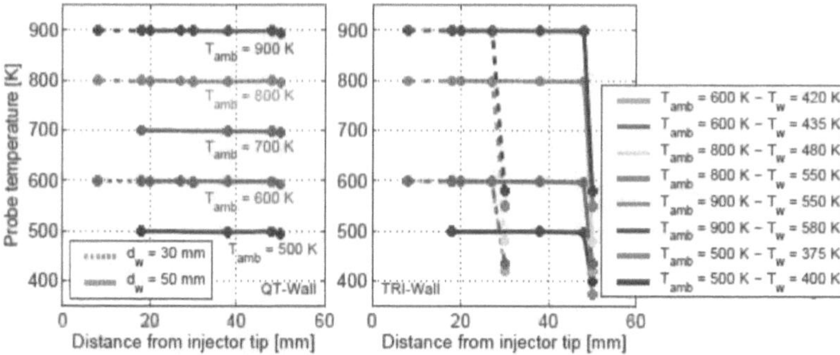

Fig. 4.8: On-axis temperatures in the test chamber. Left: Profile using the QT-Wall. Right: Profile using the TRI-Wall (the legend just applies for this right plot).

These results provide three encouraging conclusions, which give green light to go forward with the methodology: The axial distribution of gas temperature is uniform, which supports the robustness of the hardware and the approaches used in this work. Similarly, the assumption of isothermal wall at the same temperature as the surrounding gas can be made for the QT-Wall with a negligible error. Finally, the thermal boundary layer for the TRI-Wall is not thick enough to block macroscopic visualization of the spray-wall interaction and low wall temperatures can be reached with the cooling system design.

4.3 Spray images processing methodologies

The processing of all images has been done using purpose-build algorithms created in MATLAB®. The image processing methodologies implemented in this work can be divided into two main categories depending on the desired characteristic: calculation of spray geometric features, such as spray penetration or spreading along the wall for instance, and computation of temporal parameters such as start of injection (SOI), ignition delay (*ID*) or start of interaction between spray and wall (τ_w). The most common approach for any analysis is to binarize the image to, afterwards, use the contour or the intensity levels into it for calculating any specific metric.

The high-speed recorded spray movies are processed essentially following four steps per video frame. First, the image is masked to save computational time. Then, the background is subtracted from the images. After that, the image is binarized and filtered to improve the processing and finally, the contour is detected and analyzed to get the corresponding variables. On the other side, the time-averaged OH* chemiluminescence images have a different method to be processed, which will be explained in an additional subsection.

4.3.1 Image masking

An image is recognized by MATLAB® as a height-pixels × width-pixels matrix whose element values are represented by the grayscale intensity of each pixel, where 0 is black and the values are growing accordingly to the bit depth of the image (white value = 2^{bits}-1). In order to reduce processing time, non-interest pixels are omitted. This is done by creating two masks that are essentially matrices of ones or zeros, depending if the region is of interest or not. The raw image is masked by the masks by element-to-element logic product. The first mask is created from the pixel in which the injector tip is located and it is made to define the interest region with an angle. The second mask is made to suppress all pixels behind the wall. The angle of interest defined by the first mask has been changed depending on d_w and θ_w to ensure the spray is not masked in any instant of the injection event. The result of the masking process is essentially the definition of three areas where the interest region remains the same, as can be seen in the example of Figure 4.9.

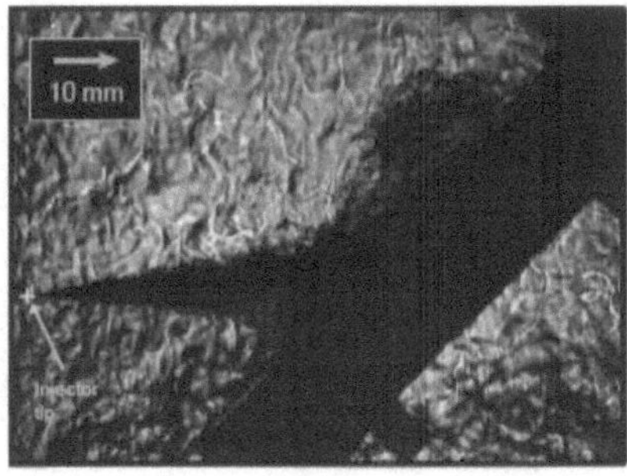

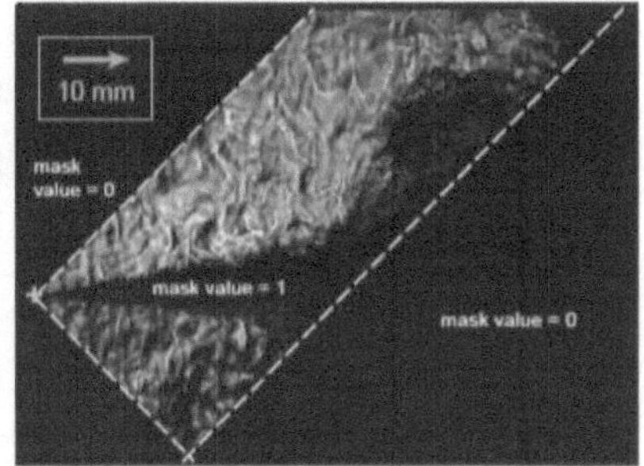

Fig. 4.9: Sample of the masking process (Wall hardware = QT-Wall; $T_{amb} = 900\,\mathrm{K}$; $\rho_{amb} = 22.8\,\mathrm{kg\,m^{-3}}$; $p_{rail} = 150\,\mathrm{MPa}$; $d_w = 50\,\mathrm{mm}$; $\theta_w = 45°$; $xO_2 = 0$; Fuel = D2; $t_{ASOI} = 2500\,\mathrm{\mu s}$). Left: Raw image. Right: Masked image (the angle of the left mask is 90°).

4.3.2 Background subtraction and binarization

The background is subtracted from the image to normalize the minimum luminosity to zero. In the DBI and natural luminosity images, where the variation between frames is mainly given by the spray, the background could be considered as static, which allows this normalization by the use of the Beer-Lambert law, presented in Equation 4.2 and an average of images without spray (first 10 images were taken).

The process is more complex for Schlieren images, where the background presents inhomogeneous-moving structures due to the density gradients of the ambient gas that is flowing into the vessel and the thermal boundary layers in solid surfaces. In this case, a new dynamic background is calculated and is constantly updated frame-to-frame to be subtracted. Two approaches have been used with this purpose:

- **Varying-background method:** Considering an image of a determined frame (Im_i), to get its corresponding background for subtraction, two regions are taken into consideration from the spray detection of the previous time step image Im_{i-1}, similarly as made in [16, 20, 38, 46, 47]. The first of them is determined by the spray pixels in Im_{i-1}, which are filled with the original background before injection. This original background can be considered as static due to the slow movement of the flow inside the vessel in comparison to spray velocities. The second region encompasses the rest of the Im_i pixels in the positions of the non-spray pixels in Im_{i-1}. Despite the steady-background assumption, one weak point of this approach is the detection of diluted or low-fuel-concentration areas of the spray, which can resemble background structures created by local strong gradients of density.

- **Temporal-derivative method:** This second method, used previously in [1, 2, 40, 48] and coupled to the previous one in [48, 49] is capable of identifying temporal variations in the spray, which is ideal to the detection of those spray areas with low fuel concentration. The pixel-wise standard deviation of the three consecutive images Im_i, Im_{i-1} and Im_{i-2} is calculated. While it is a good approach to get those diluted regions of the spray, it does not work well for regions whose intensity levels do not change in time, such as the core of the spray. In consequence, both methods are used simultaneously in a complementary way.

Those different criteria are combined in the final steps of the processing methodology with an adjustable weighted average as will be mentioned in the following subsection. The most relevant settings that can be configured for those methods are the weight assigned to each approach and a digital level threshold used to binarize the images, which is an intensity value relative to the dynamic range of the image, that represents the boundary between spray and background. The threshold calculation method is dependent of the type of images to process. In the case of Schlieren images, a fixed threshold was used with a 4 % of the dynamic range of the image, similarly as the approaches used in [2, 50]. In the case of both DBI and natural luminosity images, as discussed in [35], their pixel-wise extinction is calculated accordingly to the Beer-Lambert law (Equation 4.2), taking into account the extinction below a certain level as liquid in the case of DBI and over that value as flame in the case of NL.

4.3.3 Filtering and criteria combination

After the aforementioned steps, two binary images are obtained from both the varying-background and the temporal-derivative methods. However, the removal of the background is not perfect just by the use of the intensity threshold. In order to improve the detection, pixel erosion-dilation operations and connectivity criteria are employed to filter the images. The process is first, a two-step pixel erosion in order to eliminate background noise and small white structures. The following step is to remove all white areas below of a certain small pixel amount. To compensate the eroding strategy, a two-step dilation is applied to the white area. Finally, all remaining white islands which are

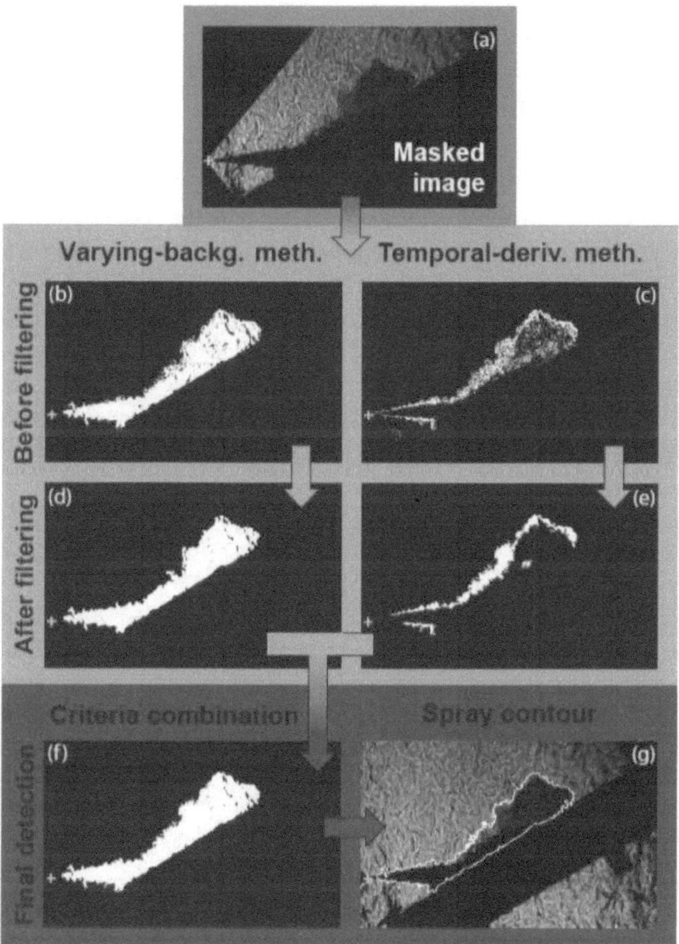

Fig. 4.10: Scheme of the processing methodology steps (Wall hardware = QT-Wall; T_{amb} = 900 K; ρ_{amb} = 22.8 kg m^{-3}; p_{rail} = 50 MPa; d_w = 30 mm; θ_w = 30°; xO_2 = 0; Fuel = D2; t_{ASOI} = 3320 µs).

not part of the spray are removed. The employed criterion was to take as spray the nearest one respect to the injector tip during the first stages of the injection and the biggest area for the rest of the movie.

Lastly, the filtered images of both processing approaches are combined in a weighted average [48], getting the binary image that represents the spray area whose boundaries define the spray contour. The whole process of spray contour detection is summarized in Figure 4.10.

4.3.4 Contour analysis

Once the spray contour is detected, post-processing is carried out to calculate the desired variables. As mentioned at the beginning of the section, there are several relevant parameters, which can be classified in two different categories: macroscopic features, related to the spray shape and geometric evolution that can be visualized and defined by the boundaries. Figure 4.11 and Figure 4.12 depict the variables that can be extracted from both side and front movies. The second type are temporal variables, that represent key times for the injection, combustion and wall impact processes. These last variables are essentially the start of injection, the ignition delay and the start of spray-wall interaction or impact time. Both ID and τ_w are reported along the document in ASOI reference.

Here below are listed the different macroscopic parameters and the criteria applied for their computation:

- **Spray penetration (S):** is obtained by selecting the furthest point of the spray contour, just taking the axial distance to the nozzle tip. This metric is just computed at free-jet conditions.
- **Spray angle (ϕ):** The plume angle is calculated as the angle between two linear fits of the spray contour data as shown in Figure 4.11 left picture. These fits are made using a data range between the 10% and the 50% of the free-jet penetration.
- **Spray spreading along the wall (Y_+):** The spray spreading along the wall is computed as the distance between the 'collision point' and the furthest contour point in the direction towards the top of the wall. This 'collision point', shown in small white horizontal marks in Figure 4.11-right, is defined as the interception between the axial spray axis and the wall plane, and not necessarily as the first point of the wall that is reached for the spray as an effect of the wall angle.
- **Spray thickness along the wall (Z_{th}):** The spray that goes along the wall has a thickness that changes with time and space. Three consecutive points from the 'collision point' (indicated in dark blue crosses in Figure 4.11 right image) have a distance of 10 mm between them, are used to measure this variable as the normal distance between the wall and the furthest point of the spray contour.

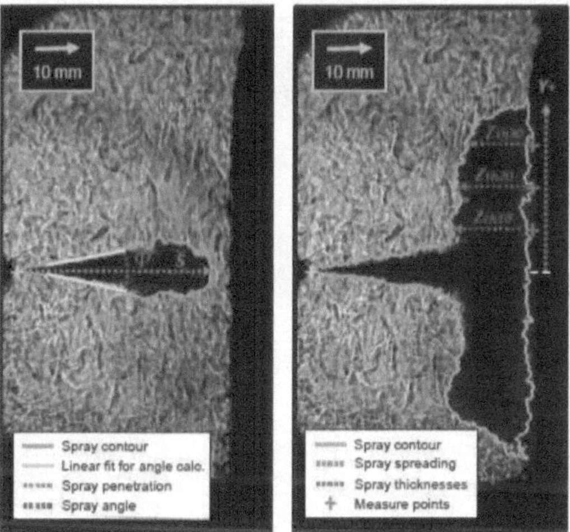

Fig. 4.11: Macroscopic parameters calculated from the side spray images. (Wall hardware = TRI-Wall; T_{amb} = 800 K; ρ_{amb} = 22.8 kg m^{-3}; p_{rail} = 150 MPa; d_w = 50 mm; θ_w = 90°; xO_2 = 0.21; Fuel = nC12). Left: Free-jet (t_{ASOI} = 430 µs). Right: Spray-wall interaction (t_{ASOI} = 2110 µs).

- **Frontal flame spreading** (X_{f+}): This characteristic just was taken from the NL images recorded by the frontal camera. From the injector tip location, that is indicated in all frames of Figure 4.12, the horizontal spreading in both left and right directions (X_{f-} and X_{f+}) is then directly calculated. As the camera is always placed horizontally, the vertical spreading of the frontal images is just a projection of the flame front displacement parallel to the wall. Taking this into account, the spreading along the wall (Y_{f-} and Y_{f+}) can be observed. Nevertheless, an important consideration is the understanding of the fact that, due to optical view reasons, this spreading is not the same as the one obtained from the side images but just a projection. Figure 4.13 shows the conceptual difference between flame Y_+ and Y_{f+}, being this last one affected by the height of the spray vortex. Therefore, flame vertical spreading is not reported from frontal images but side ones. Still, horizontal spreading can be considered reliable due to the perpendicularity between that path line with the camera, regardless of the wall angle. Analogously, the width of the flame spreading in horizontal direction (ΔX_f) can be obtained from these images.
- ***R-parameter*** (R_{par}): In agreement with models from literature [51–53], by the assumption of a conical shaped spray, the steady spray

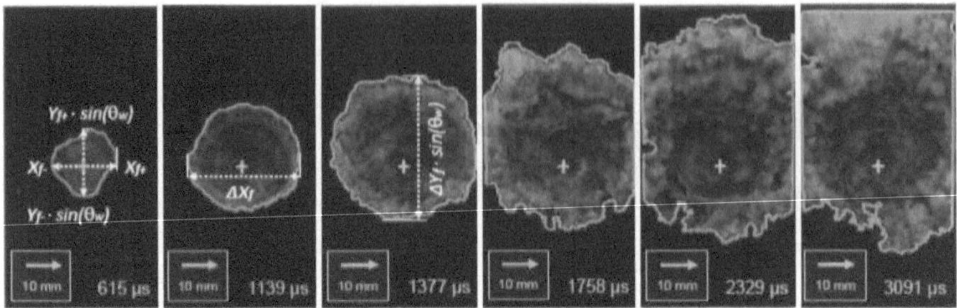

Fig. 4.12: Sequence of the flame spreading event. (Wall hardware = QT-Wall; T_{amb} = 900 K; ρ_{amb} = 22.8 kg m^{-3}; p_{rail} = 100 MPa; d_w = 50 mm; θ_w = 60°; xO_2 = 0.21; Fuel = D2). The image shows times at different temporal steps in ASOI reference.

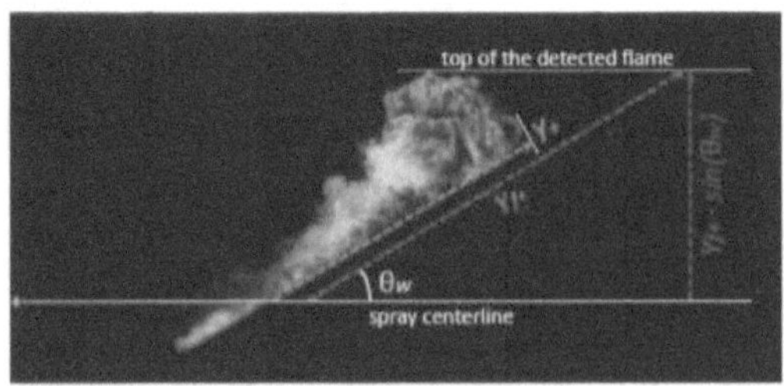

Fig. 4.13: Side view of an impinging flame and comparison between spreadings seen from the side (Y_+) and from the front (Y_{f+}). (Wall hardware = QT-Wall; T_{amb} = 800 K; ρ_{amb} = 22.8 kg m^{-3}; p_{rail} = 50 MPa; d_w = 30 mm; θ_w = 30°; xO_2 = 0.21; Fuel = D2; t_{ASOI} = 2495 µs).

vapor penetration of an inert spray is proportional to the time raised to the power of 0.5. This open the possibility to introduce a way to express the penetration and the spray development in a time-independent form and to obtain a constant parameter from spray momentum flux conservation. This variable, referred to as *R-parameter*, is defined as the derivative of the spray penetration respect to the square root of time, as represented in Equation 4.4:

$$R_{parS} = \frac{\partial S_v(t)}{\partial \sqrt{t}} \propto \dot{M}^{\frac{1}{4}} \cdot \rho_{amb}^{-\frac{1}{4}} \cdot \tan^{-\frac{1}{2}}\left(\frac{\phi_v}{2}\right) \tag{4.4}$$

Where the proportionality with the rightest part of the expression matches with several models of spray free penetration [51, 54]. *R-parameter* had proven in previous investigations to be an useful tool to study spray penetration [1, 38, 55] and spray spreading along a wall

[49]. Figure 4.14 shows an example of free vapor penetration behavior and its derivative respect to time raised to 0.5 for a different ECN injector studied in [1]. As seen in the right picture, the *R-parameter* of penetration (R_{parS}) has a constant trend with time in its steady phase with a little variability.

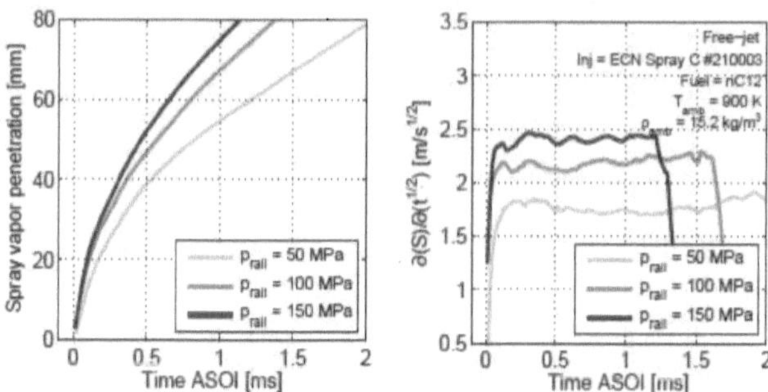

Fig. 4.14: Sample of spray penetration and its *R-parameter* at free-jet non-reactive conditions for an ECN 'Spray C' injector (adapted from [1]). Left: Vapor penetration by injection pressure. Right: *R-parameter* for same conditions.

All these geometrical variables are calculated for all repetitions that were recorded for each test conditions. Then, all this data is gathered and averaged by the use of a polynomial-average algorithm [1, 16, 56], obtaining both the shot-to-shot dispersion and the average of the signal for each time step. The penetration and *R-parameter* shown in Figure 4.14 are examples of the averaged signals obtained from 8 reps. Figure 4.15 shows both a set of different repetitions of a vapor penetration in dots and, in a black line, the average obtained by following this method. This approach is consistently used to get shot-to-shot averages from time-resolved variables along this work, as seen in Figure 4.14 for non-geometrical parameters such as Spray Intensity and its derivative. Now, in regards to the temporal parameters, the following methodologies were executed to assess to them:

- **Start of injection (*SOI*)**: The start of injection is necessary to phase results and to make proper comparisons between different points. As the optical setup allows the simultaneous recording of the same injection event with different techniques, the set of images used to calculate the SOI is the obtained from the camera set in a higher speed, whose technique uses external illumination (DBI and Schlieren), due to the impossibility of observing the start of injection in NL movies. By general rule, it was the case of the Schlieren images. In any case, the employed methodology was to take the data of the penetration curve until $S_v = 20\,\text{mm}$ and apply a numerical gridded fit (in dashed lines in Figure 4.15) based on the use of two groups of equations to pull the fitted data towards the points of the averaged penetration (black line) but taking into account partial derivatives at each side of the point to keep the fit smooth [2, 48, 56]. This fit is used to extrapolate the data and calculate the SOI as the ASOE time when $S_{fitted} = 0\,\text{mm}$.

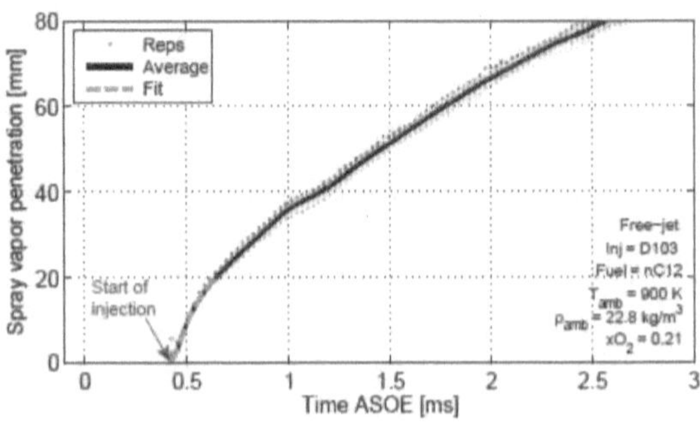

Fig. 4.15: Calculation of start of injection from a penetration curve by extrapolating a fit of the averaged data.

- **Ignition delay (*ID*)**: Through Schlieren imaging it is possible to identify two ignition stages: the start of cool flames (SoCF) or first stage of ignition, when the first indications of chemical reactions occurrence take place and, due to these changes in the mixture composition the head of the diesel spray adopts a refractive index similar to the one of the ambient gas [16, 57]; and the second stage of ignition (SSI), where high-temperature reactions occur and the spray is characterized by a rapid expansion and by the presence of high-luminosity flames. An intensity-based approach that is extensively used in free-jet diesel spray conditions [20, 25, 38] has been carried out to calculate the SSI which is considered as the ignition delay (ID) reported along the manuscript. Within the detected spray contour, the pixel-by-pixel intensity is totalized and then derived to obtain both total spray intensity and its increment as time-resolved signals (an example of those metrics in a spray-wall case is shown in Figure 4.16, where the different reps are shown in dots and the polynomial-averaged curve is shown in black lines). The local maximum value of the intensity increment is taken as ignition delay.

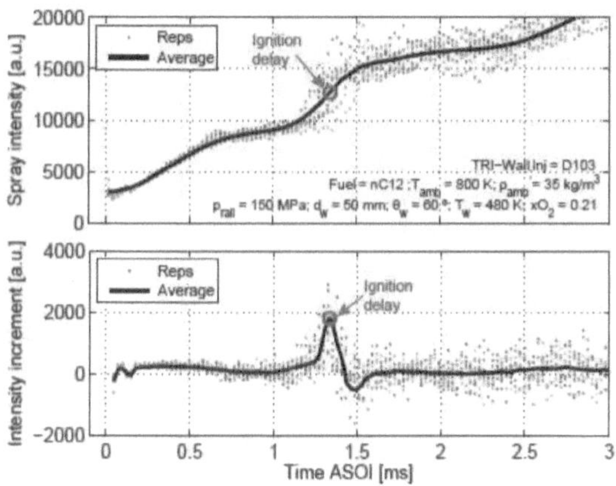

Fig. 4.16: Principle of calculation of ignition delay based on the evolution of the spray intensity and its derivative. Top: Spray intensity raw data per rep and averaged. Bottom: Spray intensity variation at the same conditions

- **Start of spray-wall interaction** (τ_w): From the definition of spray penetration, it could be considered that the start of the spray-wall interaction begins when $S_v = d_w$. Nevertheless, as can be seen in Figure 4.17, the furthest point of the spray contour detected on-spray-axis become stagnant few millimeters before reaching the wall distance. This happens because of the strong density gradients produced in the surface of the wall by the thermal boundary layer, which limits the visualization in the nearest millimeters to the wall. A characterization of the homogeneity of the gas temperature within the vessel was made before testing in order to verify the order of magnitude of this boundary layer and the feasibility of the use of Schlieren imaging in spite of it. Even when Schlieren visualization provides good quality images that allow to extract a lot of information, this is a limitation to be taken into account. To overcome it and detect the moment when the spray reaches the wall (in on-axis terms) a similar approach to the one used to calculate the start-of-injection time was used. First, from images of the wall placed in the HPHTV without heated gas flow, the real distance from the injector tip to the wall was measured (with deviations always smaller than 2 mm respect to the design d_w). Then, a numerical gridded fit was made again, with the penetration data from $0.2 \cdot d_w$ to $0.8 \cdot d_w$ to avoid largely the most transient part close to the start of the injection and the part when the spray is near to the boundary layer. This fit is then extrapolated to d_w and the time when they match is taken as τ_w.

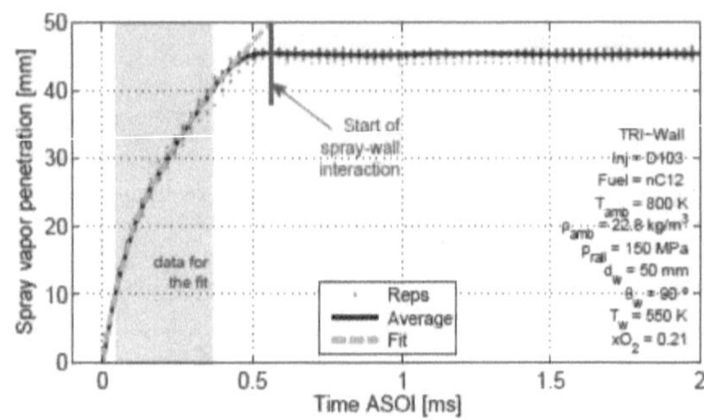

Fig. 4.17: Calculation of start of SWI from the extrapolation of the penetration curve using a numerical fit.

4.3.5 OH* chemiluminescence images processing

The OH* chemiluminescence images were processed in a systematic way following an approach based on the used by Gimeno et al. [17] for *LoL* calculation in free sprays. In the first place, the raw image (Figure 4.18-left) is divided in two through the spray axis into a top and a bottom parts. Then, a fixed threshold of 0.3 between the 5% and 95% of the max intensity level, obtaining a mask that covers the whole image but the pixels beyond this value. After that, a fixed radial distance in the detected spray is used for both bottom and top parts of the image to define a region of interest and, the location of the leftest pixel above 50% of the max intensity in the region, is taken as the *LoL*. Finally, for lift-off length calculation, both LoL_{top} and LoL_{bot} are averaged. Due to the continuous use of the intensified camera over the years, the screen where the image is formed is slightly damaged, having some black spots of burned pixels, that are presented in both Figure 4.18 and Figure 4.19. However, the camera was always aligned in order to prevent the influence of these small areas on the interest metrics and the detected contours without compromising the spatial resolution of the images.

However, a limitation of this technique is present when the wall is located in its nearest position to the injector tip (d_w = 30 mm). In this condition, the lift-off length is susceptible to be optically blocked by the spray thickness itself, as can be observed in the left picture of Figure 4.19, where the short wall distance locates the entire combustion and SWI phenomena close to the nozzle. In spite of that, there are some conditions of short lift-off length, such

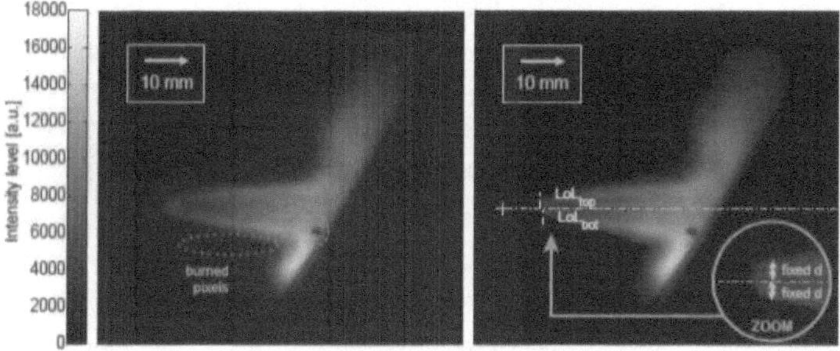

Fig. 4.18: OH* imaging sample and variables calculation in a set color scale (Wall hardware = TRI-Wall; T_{amb} = 900 K; ρ_{amb} = 35 kg m^{-3}; p_{rail} = 100 MPa; d_w = 50 mm; θ_w = 60°; T_w = 550 K; xO_2 = 0.21; Fuel = nC12). Left: Raw image. Right: Image after being masked from binarization results.

as high temperatures and densities into the chamber, that can promote the establishment of LoLs short enough to be visible for the intensify camera. One case of that is illustrated in Figure 4.19-right. This makes the lift-off length to be not accurately estimated for the entire test plan. That being said, lift-off length will be reported just for the points where it can be extracted.

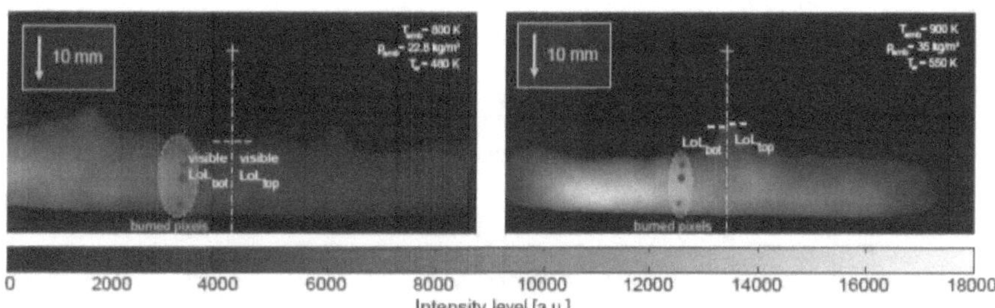

Fig. 4.19: Comparison between binarized OH* images at d_w = 30 mm and both large and short LoL conditions (Wall hardware = TRI-Wall; p_{rail} = 150 MPa; θ_w = 90°; xO_2 = 0.21; Fuel = nC12. The rest of the conditions are indicated in each picture). Left: Low temperature and density conditions. Right: High temperature and density conditions.

4.4 Wall temperature and heat flux calculation

Additionally to the image processing, the signal from the two fast thermocouples was used to obtain the surface temperature profiles at the different probe locations. The local surface temperature signals (max values around 12 mV) where directly transformed to temperature units using the correlations of Figure 3.14, resulting in a time and temperature resolutions of 10 µs (acquisition rate of 100 kHz) and around 0.206 °C, which were considered adequate for the duration of the injection and the spray-wall interaction could last approximately 3-4 ms.

All this data was input to compute the heat flux using MATLAB®. One-dimensional heat conduction within the wall was considered, making a simplification of the heat equation as follows:

$$\frac{\partial T_w}{\partial t} = \alpha_w \cdot \left(\cancelto{0}{\frac{\partial^2 T_w}{\partial x^2}} + \cancelto{0}{\frac{\partial^2 T_w}{\partial y^2}} + \frac{\partial^2 T_w}{\partial z^2} \right) \tag{4.5}$$

$$\frac{\partial T_w}{\partial t} = \alpha_w \cdot \left(\frac{\partial^2 T_w}{\partial z^2} \right) \tag{4.6}$$

Where z is the coordinate perpendicular to the surface, T_w is the temperature of the wall in any point along z, and α_w is the thermal diffusivity of 316 stainless steel, which is the material of the wall and matches with the reported by the thermocouples manufacturer ($3.476 \times 10^{-6}\,\mathrm{m^2s^{-1}}$). The assumption of 1-D heat flux is justified establishing that:

- The simulations presented in the previous chapter indicate that the normal temperature gradient is much larger in comparison to the lateral.
- The duration of the injection is too short to cause a significant radial heat flux.
- The variation of temperature in the solid wall, due to SWI takes place just in a few millimeters beyond the wall surface. This makes negligible the effect of the geometry of the holes made in the wall for the thermocouple fitting on the heat flux calculation.

Some ways to compute heat flux under this 1-D assumption, both numerical and analytical have been reviewed. In literature, there are several expressions to find heat flux analytically, like the one shown in Equation 4.7 [58, 59], where $\sqrt{k_w \rho_w c_{pw}}$ is the thermal effusivity of the thermocouple junction ($8430\,\mathrm{Ws^{0.5}m^{-2}K^{-1}}$ in this case). Nevertheless, even if this solution is fast and easy to apply and solve the mathematical problem, additional information can be extracted by other means.

$$\dot{q}_w(t) = \sqrt{\frac{k_w \rho_w c_{pw}}{\pi}} \cdot \int_{t'=-\infty}^{t} \frac{\partial T_w}{\partial t} \cdot \frac{dt'}{\sqrt{t-t'}} \tag{4.7}$$

Being said that, the approach to calculate the heat flux was to solve the numerical model presented in Figure 4.20 via finite differences method (FDM), solving the parabolic partial differential Equation 4.6 [60, 61]. Not only $\dot{q}_w(t)$ is obtained from this approach, but also temperature at any wall depth and time, assessing to the depth of the wall that is actually affected by SWI. From the known Δt of the data, the Fourier number (Fo) was kept below 0.5 for calculation stability criterion [61] accordingly to:

$$Fo = \left(\frac{\alpha_w \cdot \Delta t}{\Delta z^2}\right) < 0.5 \tag{4.8}$$

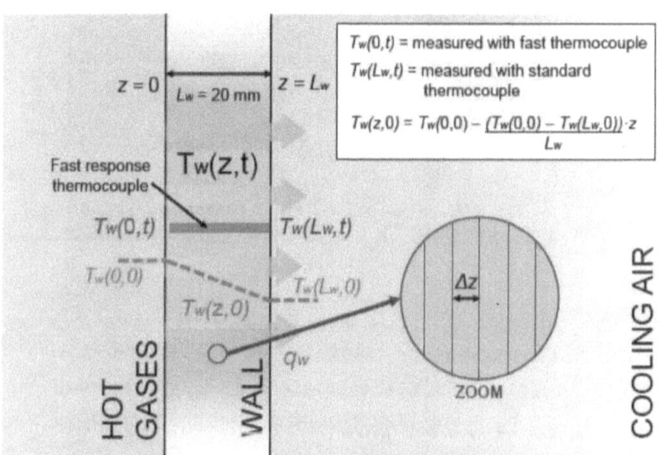

Fig. 4.20: Wall heat transfer model for heat flux calculation.

Forcing Fo to 0.3, it was possible to define a spacial resolution of $\Delta z = 11\,\mu m$, which is more than sufficient for this study. Defining $T_w(z,t)$ as function of the dimension z and the time t, the boundary conditions $T_w(0,t)$ and $T_w(L_w,t)$ are measured with the thermocouples in both the cold and hot wall faces, taking into account that the temperature of the cold face is not affected by the injection-combustion processes. From the known initial conditions $T_w(0,0)$ and $T_w(L_w,0)$, the initial temperature profile of the wall $T_w(z,0)$ is estimated as the slope between those temperatures, obtaining all the needed conditions to perform the FDM.

As a sample of this calculation, Figure 4.21 shows a map of the wall temperature variation ΔT_w respect to the initial condition $T_w(z,0)$, in order to show a representative case of how the solid wall temperature is disturbed by the SWI with time and how goes deep into the solid wall. As seen, and as happens in all cases and probes, the third aforementioned hypothesis is compliant, not having temperature variations over 3 mm of wall depth. Additionally, this fact justifies the use of a conventional thermocouple for the cooled face of the wall, located at 20 mm from the hot surface and helps to save computational time reducing the size of interest region.

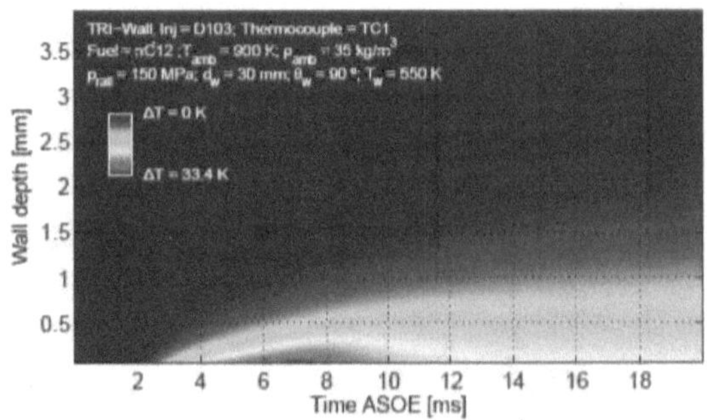

Fig. 4.21: Map of wall temperature variation respect to the initial condition $T_w(z,0)$ (Wall hardware = TRI-Wall; T_{amb} = 900 K; ρ_{amb} = 35 kg m^{-3}; p_{rail} = 150 MPa; d_w = 30 mm; θ_w = 90°; T_w = 550 K; xO_2 = 0.21; Fuel = nC12).

Finally, the calculation of the heat transfer per area unit or heat flux is made from the temperature measured in the hot surface $T_w(0,t)$ and the following temperature profile in z going through the wall, as shown in Equation 4.9, where k_w is the heat conduction coefficient of the wall:

$$\dot{q}_w(t) = k_w \cdot \left(\frac{T_w(0,t) - T_w(\Delta z,t)}{\Delta z} \right) \tag{4.9}$$

Figure 4.22 shows an example of the curves obtained with this methodology. The top plot shows the raw signal of the surface temperature obtained from the TC1 of the 90° wall of the TRI-Wall (d_{dg} = 15 mm) for one arbitrary repetition. This point is representative of the noise of the signal obtained by the two fast-response thermocouples. This signal was cleaned using a Savitzky-Golay filter [62], obtaining the dotted line. Even when some information is lost with this filter as, for instance, an underestimation of the abruptness of the temperature increasing in its beginning, this filtering step is necessary since the heat flux calculation is quite sensitive to fluctuations in the temperature signal. The variation of this temperature from the initial one is represented in Figure 4.22-middle, where it can be observed how the same rolling-average approach was employed to get a single signal from the different repetitions of the test point. Similarly, this is made for the surface heat flux (Figure 4.22-bottom). This methodology was followed for all the thermocouples.

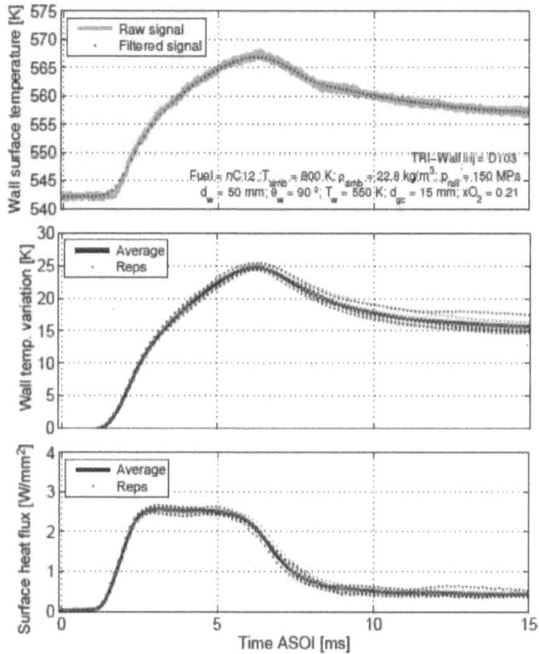

Fig. 4.22: Example of the heat flux and temperature variations obtained per rep and averaged. Top: Wall temperature variation respect to the initial condition $T_w(0,0)$. Bottom: Surface heat flux. The number of points of the reps has been fractioned by reducing the sampling frequency in order to avoid markers overlapping in the plot.

References

[1] Gimeno, Jaime, Bracho, Gabriela, Martí-Aldaraví, Pedro, and Peraza, Jesús E. "Experimental study of the injection conditions influence over

n-dodecane and diesel sprays with two ECN single-hole nozzles. Part I: Inert atmosphere". In: *Energy Conversion and Management* 126 (2016), pp. 1146–1156 (cited on pages 75, 78, 87, 91, 92).

[2] Payri, Raul, Viera, Juan Pablo, Gopalakrishnan, Venkatesh, and Szymkowicz, Patrick G. "The effect of nozzle geometry over the evaporative spray formation for three different fuels". In: *Fuel* 188 (2017), pp. 645–660 (cited on pages 75, 87, 93).

[3] Pickett, Lyle M, Genzale, Caroline L, Manin, Julien, Malbec, Louis-Marie, and Hermant, Laurent. "Measurement Uncertainty of Liquid Penetration in Evaporating Diesel Sprays". In: *ILASS Americas 23rd Annual Conference on Liquid Atomization and Spray Systems.* Ventura, CA (USA): ILASS-Americas, 2011 (cited on pages 76, 78).

[4] Payri, Raul, Garcia-Oliver, Jose Maria, Bardi, Michele, and Manin, Julien. "Fuel temperature influence on diesel sprays in inert and reacting conditions". In: *Applied Thermal Engineering* 35 (2012), pp. 185–195 (cited on page 76).

[5] Westlye, Fredrik R et al. "Penetration and combustion characterization of cavitating and non-cavitating fuel injectors under diesel engine conditions". In: *SAE Technical Paper 2016-01-0860* (2016), p. 15 (cited on page 76).

[6] Kastengren, Alan L et al. "Measurements of droplet size in shear-driven atomization using ultra-small angle x-ray scattering". In: *International Journal of Multiphase Flow* 92 (2017), pp. 131–139 (cited on page 76).

[7] Pastor, Jose Vicente, Payri, Raul, Araneo, Lucio, and Manin, Julien. "Correction method for droplet sizing by laser-induced fluorescence in a controlled test situation". In: *Optical Engineering* 48.1 (2009), p. 013601 (cited on page 76).

[8] Payri, Raul, Gimeno, Jaime, Martí-Aldaraví, Pedro, and Giraldo, Jhoan S. "Methodology for Phase Doppler Anemometry Measurements on a Multi-Hole Diesel Injector". In: *Experimental Techniques* 41.2 (2016), pp. 1–8 (cited on page 76).

[9] Kastengren, Alan L, Powell, Christopher F, Liu, Z, and Wang, J. "Time resolved, three-dimensional mass distribution of diesel sprays measured with x-ray radiography". In: *SAE Technical Paper 2009-01-0840* (2009) (cited on page 76).

[10] Pickett, Lyle M, Manin, Julien, Kastengren, Alan L, and Powell, Christopher F. "Comparison of Near-Field Structure and Growth of a Diesel Spray Using Light-Based Optical Microscopy and X-Ray Radiography". In: *SAE Technical Paper 2014-01-1412* (2014), pp. 1044–1053 (cited on page 76).

[11] Pickett, Lyle M and Siebers, Dennis L. "Soot in diesel fuel jets: effects of ambient temperature, ambient density, and injection pressure". In: *Combustion and Flame* 138.1 (2004), pp. 114–135 (cited on page 76).

[12] Payri, Raul, Gimeno, Jaime, Cardona, Santiago, and Ayyapureddi, Sridhar. "Measurement of Soot Concentration in a Prototype Multi-Hole Diesel Injector by High-Speed Color Diffused Back Illumination Technique". In: *SAE Technical Paper 2017-01-2255*. 2017 (cited on page 76).

[13] Bardi, Michele, Bruneaux, Gilles, Nicolle, André, and Colin, Olivier. "Experimental Methodology for the Understanding of Soot-Fuel Relationship in Diesel Combustion: Fuel Characterization and Surrogate Validation". In: 2017 (cited on page 76).

[14] Peterson, Brian, Reuss, David L., and Sick, Volker. "On the ignition and flame development in a spray-guided direct-injection spark-ignition engine". In: *Combustion and Flame* 161.1 (2014), pp. 240–255 (cited on page 76).

[15] Anbari Attar, Mohammadreza, Herfatmanesh, Mohammad Reza, Zhao, Hua, and Cairns, Alasdair. "Experimental investigation of direct injection charge cooling in optical GDI engine using tracer-based PLIF technique". In: *Experimental Thermal and Fluid Science* 59 (2014), pp. 96–108 (cited on page 76).

[16] Benajes, Jesus, Payri, Raul, Bardi, Michele, and Martí-aldaraví, Pedro. "Experimental characterization of diesel ignition and lift-off length using a single-hole ECN injector". In: *Applied Thermal Engineering* 58.1-2 (2013), pp. 554–563 (cited on pages 76, 79, 86, 92, 93).

[17] Gimeno, Jaime, Martí-Aldaraví, Pedro, Carreres, Marcos, and Peraza, Jesús E. "Effect of the nozzle holder on injected fuel temperature for experimental test rigs and its influence on diesel sprays". In: *International Journal of Engine Research* 19.3 (2018), pp. 374–389 (cited on pages 76, 80, 83, 95).

[18] Lequien, Guillaume, Li, Zheming, Andersson, Öivind, and Richter, Mattias. "Lift-Off Length in an Optical Heavy-Duty Diesel Engine". In: *SAE International Journal of Engines* 8.5 (2015), pp. 2015–24–2442 (cited on page 76).

[19] Rusly, Alvin M., Le, Minh K., Kook, Sanghoon, and Hawkes, Evatt R. "The shortening of lift-off length associated with jet-wall and jet-jet interaction in a small-bore optical diesel engine". In: *Fuel* 125 (2014), pp. 1–14 (cited on page 76).

[20] Payri, Raul, Viera, Juan Pablo, Pei, Yuanjiang, and Som, Sibendu. "Experimental and numerical study of lift-off length and ignition delay of a two-component diesel surrogate". In: *Fuel* 158 (2015), pp. 957–967 (cited on pages 76, 86, 93).

[21] Manin, Julien, Pickett, Lyle M, and Yasutomi, K. "Transient cavitation in transparent diesel injectors". In: *ICLASS 14th Triennial International Conference on Liquid Atomization and Spray Systems.* Chicago, 2018, pp. 1–9 (cited on page 76).

[22] Viera, Juan Pablo et al. "Linking instantaneous rate of injection to X-ray needle lift measurements for a direct-acting piezoelectric injector". In: *Energy Conversion and Management* 112 (2016), pp. 350–358 (cited on page 76).

[23] Duke, Daniel J et al. "Internal and near nozzle measurements of Engine Combustion Network 'Spray G' gasoline direct injectors". In: *Experimental Thermal and Fluid Science* 88 (2017), pp. 608–621 (cited on page 76).

[24] Jung, Yongjin, Manin, Julien, Skeen, Scott A, and Pickett, Lyle M. "Measurement of Liquid and Vapor Penetration of Diesel Sprays with a Variation in Spreading Angle". In: *SAE Technical Paper 2015-01-0946* (2015) (cited on pages 76, 78).

[25] Payri, Raul, Salvador, Francisco Javier, Manin, Julien, and Viera, Alberto. "Diesel ignition delay and lift-off length through different methodologies using a multi-hole injector". In: *Applied Energy* 162 (2016), pp. 541–550 (cited on pages 76, 79, 93).

[26] Pastor, Jose Vicente, Payri, Raul, Garcia-Oliver, Jose Maria, and Briceño, Francisco Javier. "Schlieren Methodology for the Analysis of Transient Diesel Flame Evolution". In: *SAE International Journal of Engines* 6.3 (2013), pp. 1661–1676 (cited on page 76).

[27] Montanaro, Alessandro et al. "Schlieren and Mie Scattering Visualization for Single- Hole Diesel Injector under Vaporizing Conditions with Numerical Validation". In: *SAE Technical Paper* (2014) (cited on page 76).

[28] Lillo, Peter M, Pickett, Lyle M, Persson, Helena, Andersson, Öivind, and Kook, Sanghoon. "Diesel Spray Ignition Detection and Spatial/Temporal Correction". In: *SAE Technical Paper 2012-01-1239* (2012) (cited on page 76).

[29] Allocca, Luigi, Lazzaro, Maurizio, Meccariello, G., and Montanaro, Alessandro. "Schlieren visualization of a GDI spray impacting on a heated wall: Non-vaporizing and vaporizing evolutions". In: *Energy* 108 (2016), pp. 93–98 (cited on page 76).

[30] Gawthrop, D. B. "Applications of the Schlieren method of photography". In: *Review of Scientific Instruments* 2.9 (1931), pp. 522–531 (cited on page 76).

[31] Gladstone, J H and Dale, T P. "Researches on the Refraction, Dispersion, and Sensitiveness of Liquids". In: *Philosophical Transactions of the Royal Society of London* 153 (1863), pp. 317–343 (cited on page 76).

[32] Settles, Gary S. *Schlieren and Shadowgraph Techniques*. Berlin, Heidelberg: Springer Berlin Heidelberg, 2001, p. 376 (cited on page 77).

[33] Manin, Julien, Bardi, Michele, and Pickett, Lyle M. "Evaluation of the liquid length via diffused back-illumination imaging in vaporizing diesel sprays". In: *Comodia*. Fukuoka, 2012 (cited on page 77).

[34] Westlye, Fredrik R et al. "Diffuse back-illumination setup for high temporally resolved extinction imaging". In: *Applied Optics* 56.17 (2017) (cited on page 78).

[35] Payri, Raul, Salvador, Francisco Javier, Marti-Aldaravi, Pedro, and Vaqucrizo, Daniel. "ECN Spray G external spray visualization and spray collapse description through penetration and morphology analysis". In: *Applied Thermal Engineering* 112 (2017), pp. 304–316 (cited on pages 78, 87).

[36] Ghandhi, J B and Heim, D M. "An optimized optical system for backlit imaging". In: *Review of Scientific Instruments* 80 (2009) (cited on pages 78, 79).

[37] Gaydon, A G. *The Spectroscopy of Flames*. Springer Netherlands, 1974 (cited on page 78).

[38] Payri, Raul, Salvador, Francisco Javier, Gimeno, Jaime, and Peraza, Jesús E. "Experimental study of the injection conditions influence over n-dodecane and diesel sprays with two ECN single-hole nozzles. Part II: Reactive atmosphere". In: *Energy Conversion and Management* 126 (2016), pp. 1157–1167 (cited on pages 78, 86, 91, 93).

[39] Pittermann, Roland. "Spectroscopic Analysis of the Combustion in Diesel and Gas Engines". In: *MTZ worldwide* 69.7 (2008), pp. 66–73 (cited on pages 78, 79).

[40] Bardi, Michele et al. "Engine Combustion Network: Comparison of Spray Development, Vaporization, and Combustion in Different Combustion Vessels". In: *Atomization and Sprays* 22.10 (2012), pp. 807–842 (cited on pages 79, 87).

[41] Peters, Norbert. *Turbulent Combustion.* Cambridge Monographs on Mechanics. Cambridge University Press, 2000 (cited on page 79).

[42] Higgins, Brian and Siebers, Dennis L. "Measurement of the Flame Lift-Off Location on DI Diesel Sprays Using OH Chemiluminescence". In: *SAE Paper 2001-01-0918* (2001) (cited on pages 79, 80).

[43] Bardi, Michele, Bruneaux, Gilles, and Malbec, Louis-Marie. "Study of ECN Injectors' Behavior Repeatability with Focus on Aging Effect and Soot Fluctuations". In: *SAE Technical Paper 2016-01-0845* (2016) (cited on page 80).

[44] Pickett, Lyle M, Siebers, Dennis L, and Idicheria, Cherian A. "Relationship Between Ignition Processes and the Lift-Off Length of Diesel Fuel Jets". In: *SAE Paper 2005-01-3843* 724 (2005) (cited on page 80).

[45] Chartier, Clément, Aronsson, Ulf, Andersson, Öivind, Egnell, Rolf, and Johansson, Bengt. "Influence of jet-jet interactions on the lift-off length in an optical heavy-duty DI diesel engine". In: *Fuel* 112 (2013), pp. 311–318 (cited on page 80).

[46] Payri, Raul, Gimeno, Jaime, Viera, Juan Pablo, and Plazas, Alejandro Hernan. "Schlieren visualization of transient vapor penetration and spreading angle of a prototype diesel direct-acting piezoelectric injector". In: *ICLASS 2012.* 2012, pp. 1–8 (cited on page 86).

[47] Payri, Raul, Gimeno, Jaime, Bracho, Gabriela, and Vaquerizo, Daniel. "Study of liquid and vapor phase behavior on Diesel sprays for heavy duty engine nozzles". In: *Applied Thermal Engineering* 107 (2016), pp. 365–378 (cited on page 86).

[48] Payri, Raul, Salvador, Francisco Javier, Bracho, Gabriela, and Viera, Alberto. "Differences between single and double-pass schlieren imaging on diesel vapor spray characteristics". In: *Applied Thermal Engineering* 125 (2017), pp. 220–231 (cited on pages 87, 88, 93).

[49] Payri, Raul, Gimeno, Jaime, Peraza, Jesús E., and Bazyn, Tim. "Spray / wall interaction analysis on an ECN single-hole injector at diesel-like conditions through Schlieren visualization". In: *Proc. 28th ILASS-Europe, Valencia* September (2017) (cited on pages 87, 92).

[50] Payri, Raul, Viera, Juan Pablo, Gopalakrishnan, Venkatesh, and Szymkowicz, Patrick G. "The effect of nozzle geometry over ignition delay and flame lift-off of reacting direct-injection sprays for three different fuels". In: *Fuel* 199 (2017), pp. 76–90 (cited on page 87).

[51] Naber, Jeffrey D and Siebers, Dennis L. "Effects of Gas Density and Vaporization on Penetration and Dispersion of Diesel Sprays". In: *SAE Paper 960034* (1996) (cited on pages 90, 91).

[52] Araneo, Lucio, Coghe, Aldo, Brunello, G, and Cossali, Gianpietro E. "Experimental Investigation of gas density effects on diesel spray penetration and entrainment". In: *SAE Paper 1999-01-0525* (1999) (cited on page 90).

[53] Desantes, Jose Maria, Payri, Raul, Salvador, Francisco Javier, and Gil, Antonio. "Development and validation of a theoretical model for diesel spray penetration". In: *Fuel* 85.7-8 (2006), pp. 910–917 (cited on page 90).

[54] Gimeno, Jaime. "Desarrollo y aplicación de la medida de flujo de cantidad de movimiento de un chorro Diesel". PhD book. E.T.S. Ingenieros Industriales. Universitat Politècnica de València, 2008 (cited on page 91).

[55] Martí-Aldaraví, Pedro. "Development of a computational model for a simultaneous simulation of internal flow and spray break-up of the Diesel injection process". PhD book. Valencia: Universitat Politècnica de València, 2014 (cited on page 91).

[56] Payri, Raul, Salvador, Francisco Javier, Gimeno, Jaime, and Viera, Juan Pablo. "Experimental analysis on the influence of nozzle geometry over the dispersion of liquid n-dodecane sprays". In: *Frontiers in Mechanical Engineering* 1 (2015), pp. 1–10 (cited on pages 92, 93).

[57] Bardi, Michele. "Partial needle lift and injection rate shape effect on the formation and combustion of the Diesel spray". PhD book. Valencia (Spain): Universitat Politècnica de València, 2014 (cited on page 93).

[58] Moussou, Julien, Pilla, Guillaume, Rabeau, Fabien, Sotton, Julien, and Bellenoue, Marc. "High-frequency wall heat flux measurement during wall impingement of a diffusion flame". In: *International Journal of Engine Research* (2019) (cited on page 97).

[59] Meingast, Ulrich, Staudt, Michael, Reichelt, Lars, and Renz, Ulrich. "Analysis of Spray / Wall Interaction Under Diesel Engine Conditions". In: *SAE Technical Paper 2000-01-0272* 724 (2000), pp. 1–15 (cited on page 97).

[60] Carnahan, Brice, Luther, H. A., and Wilkes, James O. *Applied Numerical Methods.* Wiley, 1969 (cited on page 98).

[61] Nakamura, Joyce. *Applied Numerical Methods with Software.* 1st. Upper Saddle River, NJ, USA: Prentice Hall PTR, 1990 (cited on page 98).

[62] Savitzky, Abraham. and Golay, M. J. E. "Smoothing and Differentiation of Data by Simplified Least Squares Procedures." In: *Analytical Chemistry* 36.8 (1964), pp. 1627–1639 (cited on page 100).

Chapter 5

Evaporative inert spray impingement

A first and simplified approach to study spray-wall interaction processes was carried out by spray visualization at inert conditions. In this situation, the spray has the coexistence of both liquid and vapor phases due to its atomization and complete vaporization while it is advancing downstream, as a result of both high density and high temperature conditions of the vessel. Diesel self-ignition was suppressed by filling the ambient atmosphere entirely with nitrogen ($xN_2 = 1;\ \ xO_2 = 0$). Despite being a simpler approximation respect to the reactive case, an adequate good prediction of vapor penetration has proven to be a valuable starting point to refine reactive penetration and fuel mixture fraction [1–5]. Additionally, liquid length measurements are a reference frame to the understanding of the evaporation regime of the spray [6–10]. Similar data about the macroscopic characteristics of the spray, which is widely extracted for free-jet in literature [5, 11–14] is presented for impinging sprays.

The test plan, presented in Table 5.1, is focused on ECN nominal conditions and some other has been added to evaluate parametric variations around them. The hardware employed for this campaign is the quartz transparent wall and also, free-jet conditions were tested as reference. Therefore, a set of wall positions is also proposed in the test matrix. Furthermore, two different fuels were tested. A rail pressure of 200 MPa was used for commercial diesel #2, but due to its lower freezing point [15, 16], n-dodecane tests were carried out with a max injection pressure of 150 MPa, in order to prevent fuel freezing and clogging in the high-pressure lines.

Table 5.1: Inert impinging spray test plan.

Variable	Values	Units
Wall hardware	None (free-jet) - QT-Wall	-
Fuel	nC12$^{\triangledown}$ - D2	-
Injector	Bosch 3-22 ECN Spray D (D103)	-
Energizing time (ET)	2.5	ms
Tip temperature (T_{tip})	363	K
Oxygen fraction (xO_2)	0	-
Gas temperature (T_{amb})	700$^{\triangledown}$ - 800 - 900	K
Gas density (ρ_{amb})	22.8 - 35	kg/m^3
Injection pressure (p_{rail})	50 - 100 - 150 - 200$^{\triangle}$	MPa
Wall distance (d_w)	30$^{\triangledown}$ - 50	mm
Wall angle (θ_w)	30 - 45$^{\otimes\triangle}$ - 90	$^{\circ}$
Total test points	196	points

$^{\triangledown}$ Not all possible combinations with other variables were tested
$^{\triangle}$ Only for diesel (D2) tests.
$^{\otimes}$ Only for d_w = 50 mm. Please refer to Figure 3.5-right

Similarly, the optical setup employed is the one illustrated in Figure 4.4. Some details of this optical arrangement are summarized in Table 5.2. The LED pulse for DBI is synchronized with the injector energizing and the camera triggers, and has a duration of 57 ns. Due to the absence of flame luminosity the use of filters was not necessary, which allowed to reduce the exposure time of the cameras to the minimum, still obtaining sharp and high-quality images.

Table 5.2: Details of the optical setup by technique (Figure 4.4).

Feature	Diffused back-illumination	Schlieren imaging
Camera	Photron SA5	Photron SA-X2
Sensor type	CMOS	CMOS
LED pulse duration	57 ns	-
Diaphragm gap diam.	-	4 mm
Frame rate	25 kfps	40 kfps
Shutter time	1.00 μs	1.01 μs
Pixels/mm ratio	7.00	5.88
Reps per point	8	8

5.1 Vapor spray in nozzle-wall region

In Figure 5.1, evolution with time of spray vapor penetration at free-jet conditions can be seen for different wall configurations and fuels. In the case of the tests with a wall, penetration is observed until impact takes place. In the lowest plots, the *R-parameter* for this vapor penetration (its derivative respect to the square root of time) is shown for the same tests. Ambient density, temperature and the injection pressure were arbitrarily selected to show the trends, which are similar for all different conditions. Same similarity can be found in the order of standard deviation, which is represented with a lighter shade behind the mean penetration curve. The first point that can be highlighted is how, for fixed operating conditions and fuels, all the tests have very similar penetrations, in terms of penetration itself and its *R-parameter*, which indicates that the momentum flux is similar for all those cases. Even when this may be evident, it supports the results consistency and the right control of test conditions. Furthermore, it is verified that the wall does not affect the continuous measurement of the experimental conditions respect to the free-jet, being intrusive just in the desired level, allowing the observation and analysis of spray-wall interaction.

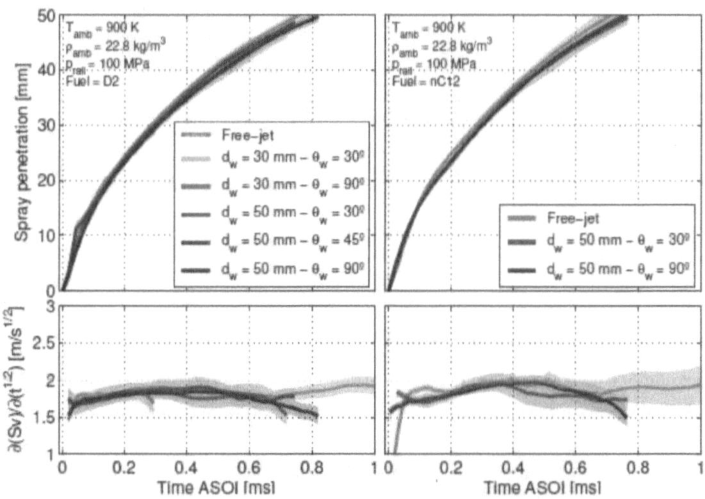

Fig. 5.1: Free vapor penetrations (top) and their *R-parameter* (bottom) for different wall configurations and fuels ($T_{amb} = 900\,\mathrm{K}$; $\rho_{amb} = 22.8\,\mathrm{kg\,m^{-3}}$; $p_{rail} = 100\,\mathrm{MPa}$). Left: Results with diesel #2. Right: n-dodecane sprays

Comparing right and left sets of plots in Figure 5.1, it can be observed how n-dodecane curves are consistently faster than diesel ones. Several free penetration models describe a noticeable influence of internal flow parameters, such as nozzle outlet diameter (D_o), area coefficient (C_a) and velocity coefficient (C_v) on penetration (S_v) [1, 12, 14, 17]. For instance, the model proposed by Naber and Siebers [12] indicates that $S_v \propto C_a^{0.25} \cdot C_v^{0.5} \cdot D_o^{0.5}$. Similarly, as the outlet diameter is not affected by the fuel, it happens for C_a, which is not influenced by fuel properties due to the non-cavitating nature of this nozzle that avoids effective orifice area contractions. Nevertheless, for the same operation conditions, C_v increases with Reynolds number (Re), taking into account that $Re \propto \mu_f^{-1}$, where μ_f is the dynamic viscosity is affected by the fluid properties. The fact of n-dodecane being less viscous than diesel #2 confers to it a higher C_v, providing an explanation of the faster nC12 spray.

Figure 5.2 depicts the behavior of free vapor penetration for different injection pressures and temperatures, for a random wall configuration (d_w = 50 mm; θ_w = 90°). As expected, rail pressure has a strong impact on spray velocity since the very beginning of the injection event due to the increase of the momentum flux at higher p_{rail}. Additionally, it can be seen how this effect is shortened when both pressure and aerodynamic drag are higher. On the other hand, gas temperature does not show important effects on vapor tip penetration. Those results are congruent with literature [1, 2, 18]. Figure 5.3 shows the effect of ambient density together with the one of fuel property

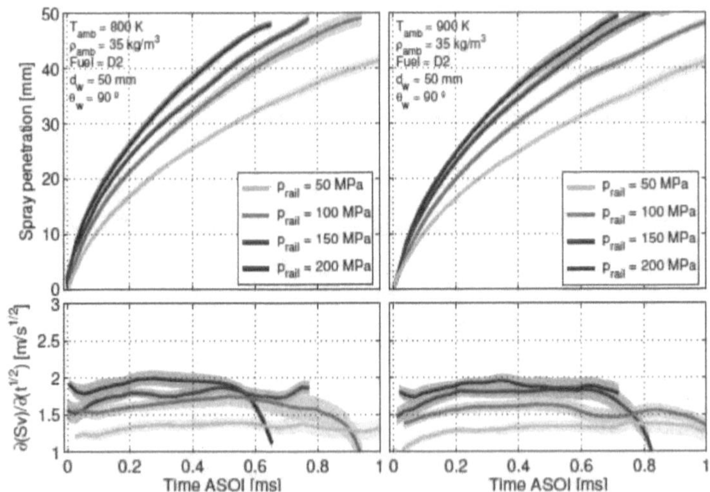

Fig. 5.2: Free vapor penetrations (top) and their *R-parameter* (bottom) for different injection pressures and ambient temperatures (ρ_{amb} = 35 kg m^{-3}; d_w = 50 mm; θ_w = 90°; Fuel = D2). Left: Test conditions at T_{amb} = 800 K. Right: Gas temperature T_{amb} = 900 K

variation, for a free-jet configuration. The effect of density appears when the effects of gas entrainment and momentum transfer between the gas and the spray are significant for spray development. This influence of density is clear: the higher the density, the slower the spray advancement into the chamber. Moreover, the effect of fuel properties on spray penetration is consistent with Figure 5.1 and described above.

τ_w is the name given to the start of spray-wall interaction, defined from the SOI. Some of the results for τ_w are shown in Figure 5.4 for a fixed wall angle and fuel. As it could be expected, it depends entirely on the penetration behavior, which makes start of SWI to take place quicker for higher injection pressures and lower gas densities, and to be negligibly influenced by ambient temperature. The longer delay that the spray takes to reach the wall depending on the wall distance . Both variations of τ_w with injector-wall distance and ambient density seem to be progressively reduced at high rail pressures. Figure 5.5 illustrates the effect of wall conditions and fuel properties. For a fixed d_w, wall inclination angle has not an important effect on τ_w, as it could be assumed from penetration curves. There is a noticeable effect of the fuel properties (shorter τ_w for n-dodecane, which accordingly to Figures 5.1 and 5.2, generates a faster spray), which is weaker than the

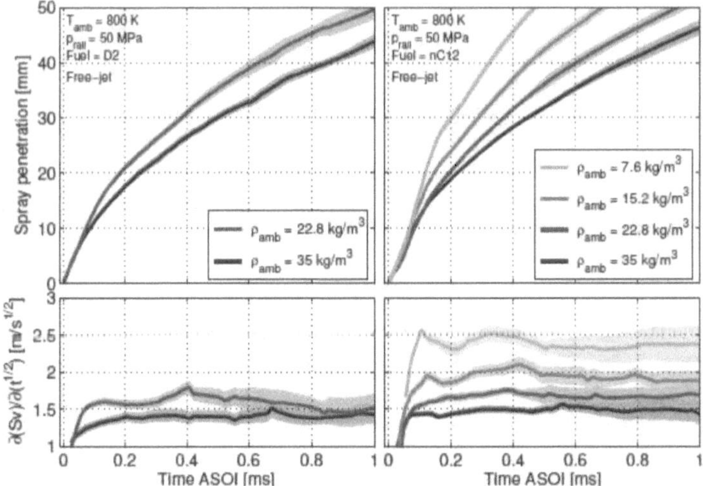

Fig. 5.3: Free vapor penetrations (top) and their *R-parameter* (bottom) for different ambient densities and fuels (T_{amb} = 800 K; p_{rail} = 50 MPa; free-jet). Left: penetrations with diesel #2. Right: Results with n-dodecane

produced by density or injection pressure. This effect and even the one of the wall distance, also seem to be less significant as injection pressure is increased.

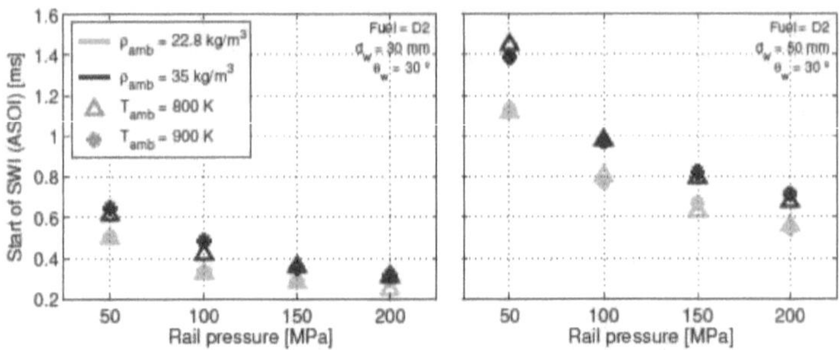

Fig. 5.4: Start of spray-wall interaction calculated for different injection pressures and ambient conditions (θ_w = 30°; Fuel = D2). Left: τ_w at d_w = 30 mm. Right: Results for d_w = 50 mm

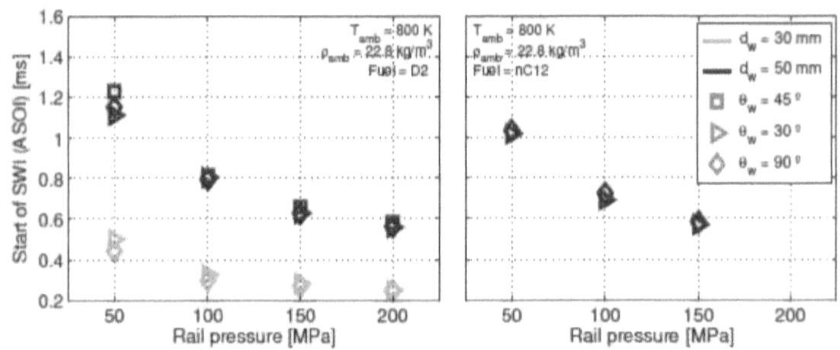

Fig. 5.5: Start of spray-wall interaction calculated for different injection pressures and wall conditions (T_{amb} = 800 K; ρ_{amb} = 22.8 kg m^{-3}). Left: Results for diesel #2. Right: SWI delay for n-dodecane.

A comparison between tests at different ambient densities and temperatures in terms of vapor spray angle evolution in time, is shown in Figure 5.6. The selected criteria to compute angle, along with the contour detection methodology, allow to determine steady angles with a very low fluctuation and uncertainty. As seen, the higher gas entrainment rate within the spray at more dense environments widens the spray angle. Ambient temperature appears to have a negligible effect on the spray angle. This happens accordingly to literature [12, 19, 20], since both fluids are low viscosity fuels with $(\rho_f/\rho_{amb}) \cdot (Re_f/We_f)^2 > 1$ (this calculation is over 8 for all test points). Left plot shows tests at free-jet conditions while right plot has SWI cases, showing two aspects that are similar for the rest of the test conditions too: there is not an evident temporal variation in the spray angle behavior when spray-wall interaction starts, and effects from varying temperature or density are equivalent for both free-jet and SWI conditions. This stabilization in the spray angle despite having a wall inside the vessel, allows to calculate a mean value using an averaging window that goes from 1 ms to 4 ms ASOI for all tests, taking into account that the entire test plan has the same energizing time and very similar start-of-injection (no longer differences than 0.12 ms).

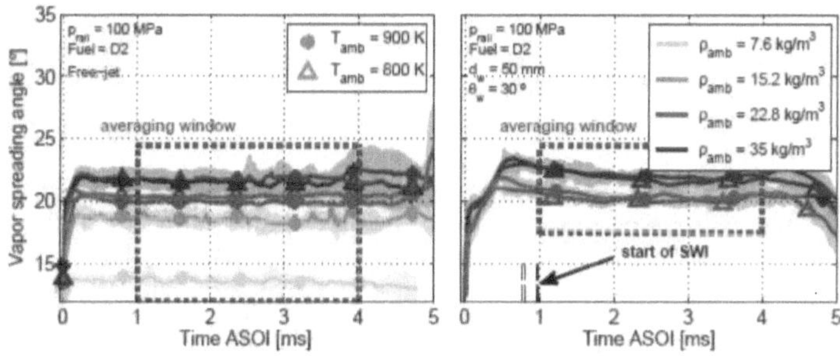

Fig. 5.6: Spray angle vs. time ASOI at different ambient densities and temperatures (p_{rail} = 100 MPa; Fuel = D2). Left: Results for the free-jet configuration. Right: Wall arrangement (d_w = 50 mm; θ_w = 30°)

Some results of these averaged spray angles are shown in Figure 5.7, whose left plot illustrates the trends for the free-jet case with different operating conditions and fuels; while the right one shows different densities and wall configurations. As seen in Figure 5.6, density is the strongest factor, whose effect is shortened when its value is greater. In the left plot it can be seen how the effect of injection pressure is not clear and it is practically negligible. On the other hand, an influence to be considered is the one of the fuel properties over the spray angle, which is weaker than the effect of density, but similarly it is more significant at lower spreading angles. As previously presented in [14], the greater viscosity and smaller density ratio (ρ_{amb}/ρ_f) of a D2 spray tend to attribute a narrower spreading angle to it.

In the right plot of Figure 5.7 it is observable how for the same chamber and injection conditions, there is not a significant effect of the wall configuration in terms of distance or angle over the steady spray angle. Nevertheless, even while it is still within the experimental uncertainty, free-jet tests show a slightly narrower steady angle than the impinging sprays, which is consistent for all tests. This trend could be addressed in a large extent to the gas entrainment capabilities of the spray. Figure 5.8 shows a set of spray contours

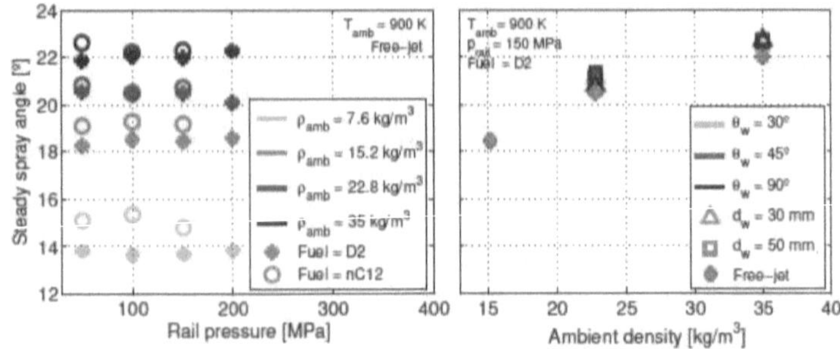

Fig. 5.7: Mean spray angle of different test points. Left: Angle variation with gas thermodynamic conditions, injection pressures and fuels (T_{amb} =900 K; free-jet). Right: Effect of wall configuration (T_{amb} =900 K; p_{rail} = 100 MPa; Fuel = D2)

at wall and free-jet conditions and three different times after start of injection, which agrees with the SWI vs free-jet spray angle trend and serves to illustrate it. While before impingement, both sprays have a similar angle and a penetration slightly larger for the free-jet, after the impact the impinging spray is spread along the wall in all directions, and the surface portion that is directly exposed to the ambient gases is reduced respect to the free-jet

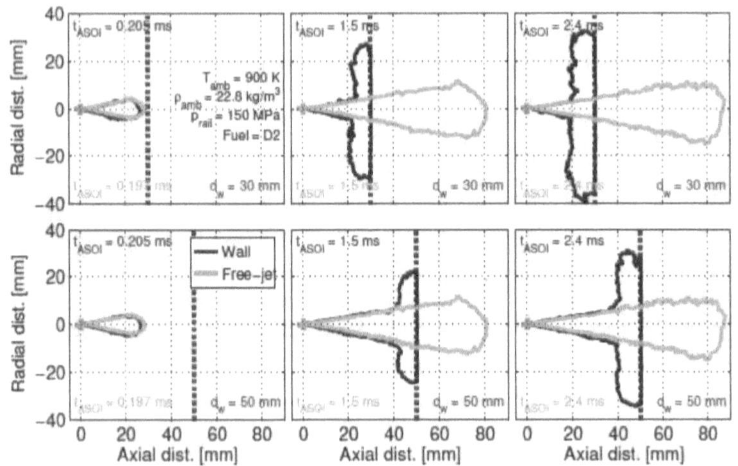

Fig. 5.8: Comparison between SWI and free-jet contours at different times (T_{amb} =900 K; free-jet). Right: Effect of wall configuration (T_{amb} = 900 K; ρ_{amb} = 35 kg m^{-3}; p_{rail} = 150 MPa; Fuel = D2). Top set: Free-jet vs. 30 mm - 90° wall. Bottom set: Free-jet vs. 50 mm - 90° wall

due to the contact with the wall face. However, when the spray spreads more along the wall, and it is further from the spray axis, the surface in contact with the surrounding gases increases in a greater rate than the free-jet case and it becomes larger, allowing more gas entrainment into the spray. Additionally, the change of the main direction of the spray momentum for different orientations drags the spray in those directions, resulting into a widening of the spray angle. Figure 5.8 also serves well as an introduction of the spray-wall interaction visualization and its macroscopic characteristics and how the spreading and thicknesses of the spray are affected by the wall distance.

To go further with the validation of the hypothesis the effect of gas-spray contact area on the angle, the former was estimated for ideally axisymmetric sprays (condition nearly seen in free-jet and SWI situations at $\theta_w = 90°$) by using the Pappus theorem to get surface areas of revolution solids from the curves defined by their contour. The total area calculated is the sum of the one obtained by revolving π radians both the bottom and top halves of the spray contours. In the case of the spray-wall interaction tests, the spray surface portion which in contact with the wall was subtracted to the total surface, in order to compute just the area that is in direct contact with the surrounding gases. Figure 5.9 depicts the results of this calculation, showing the same conditions illustrated in Figure 5.8 in its left plot. As seen, while the area for the free-jet constantly increments until the spray reaches the optical limit for SWI cases, the growth rate of spray-gas contact area is faster and it is gradually incremented. The right plot shows a slower spray (same conditions but at an injection pressure of 50 MPa) exhibiting this trend in a more evident way, where the slope of the spray-gas contact surface increases

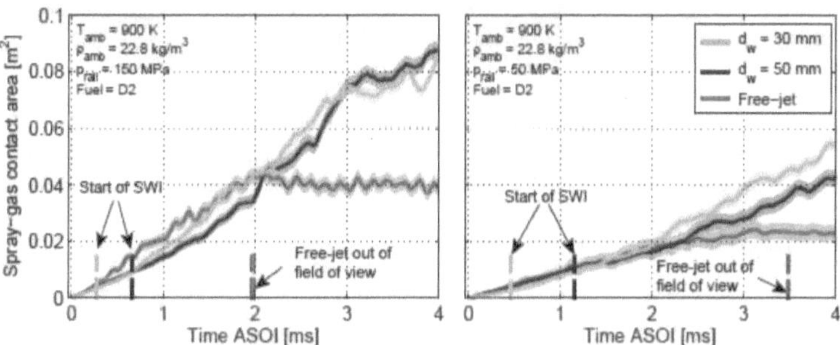

Fig. 5.9: Estimated surface area of the spray that is exposed to the in-chamber gas, compared for ideally axisymmetric sprays (free-jet and sprays against $\theta_w = 90°$ walls). Left: $p_{rail} = 150$ MPa. Right: $p_{rail} = 50$ MPa

during spray-wall interaction, which enhances the gas entrainment into the spray.

5.2 Vapor SWI visualization

The previous analyses have been made by considering parameters that are observed downstream from the wall and that, even when some of them could be influenced by SWI as it is the case of spray cone angle, its definition is not given by the presence of a wall. Hereinafter, metrics defined after the spray wall interaction are discussed. At this point, Figure 5.10 is shown in order to have a sample of the images that were observed by the use of the Schlieren technique and their detected contours. At the top, three images of an inclined wall $\theta_w = 45°$ are illustrated while at the bottom, a perpendicular wall case at the same operating conditions is found.

Fig. 5.10: Samples of a random repetition of the images observed via Schlieren imaging at inert conditions (T_{amb} = 800 K; ρ_{amb} = 22.8 kg m^{-3}; p_{rail} = 200 MPa; d_w = 50 mm; Fuel = D2). Top set: Wall inclined $\theta_w = 45°$. Bottom set: Wall angles set at $\theta_w = 90°$

The next three figures show the spreading of the spray in the same direction of wall inclination (wall top) vs. time defined after the beginning of the spray impact with the wall or τ_w. It is important to highlight that this time reference is employed in the plots to achieve a better understanding of the spray-wall interaction from its beginning, regardless of the conditions before the spray reaches the wall. Nevertheless, from the point of view of the whole injection process, it is also interesting to consider the effect of different parameters variations on the spray development and its advancement into the chamber in an ASOI reference, in order to understand how atomization and gas-fuel mixing can be enhanced in synchronized injections by changing, for instance the piston geometry or the SWI timing. Therefore, this difference with phases will also be discussed.

As it can be appreciated in Figure 5.11, the trends of spray spreading with injection and ambient conditions are analogous to the ones previously found for free penetration. Injection pressure produces a faster spray with a larger spreading along the wall, while the gas interaction with the spray tends to decelerate it in a stronger extent at higher ambient densities. Gas temperature tendency to have no effect on inert spray evolution seems to remain in spray-wall interaction conditions, taking into account that wall temperature is nearly the same as T_{amb}, which makes thermal spray-wall interaction similar to the one between the spray and its surroundings. It is important to highlight that, since these behaviors match with free-jet penetration ones and spray promptness to reach the wall, differences between spreadings at different conditions are even larger under an ASOI reference than 'after spray-wall interaction start' (which is the reference shown in Figures

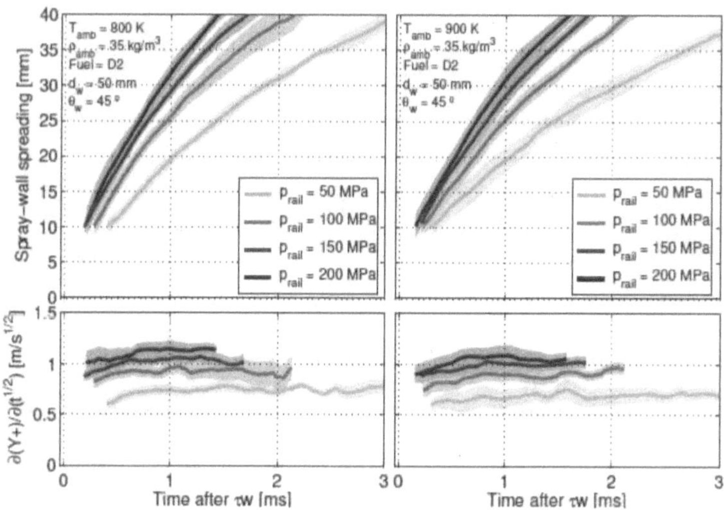

Fig. 5.11: Spreading along the wall (top) and its *R-parameter* (bottom) for different injection pressures and temperatures (ρ_{amb} = 35 kg m^{-3}; d_w = 50 mm; θ_w = 45°; Fuel = D2). Left: Test conditions at T_{amb} = 800 K. Right: Gas temperature T_{amb} = 900 K

5.11, 5.12 and 5.13). *R-parameter*, which is not affected by this time phase, not only confirms the tendencies explained above, but also proves that, similarly as free-jet spray evolution, its spreading along the wall is also proportional to square root of time. In other words, even when spray-wall impact represents spray deposition, momentum losses and its distribution in all directions along the wall and, regardless of the continuous wall contact with the spray after τ_w, the plate does not affect the nature of the spray advancement in a pressurized atmosphere. Figure 5.12 also shows equivalence between free-jet and SWI situations in terms of trends while varying gas density and fuel. nC12 sprays seem to have a slightly faster spreading than D2. Nevertheless, differences in fuel viscosities do not seem to be relevant in vapor spreading along a solid surface. Ambient density, as expected, decelerates spray advancement due to gas entrainment effects.

On the other hand, Figure 5.13 shows spreading at different wall positions. At $\theta_w = 90°$, the free spray momentum is distributed similarly in all directions along the wall after collision takes place. However, for inclined walls, this distribution has the same predominant direction in which Y_+ is measured, while the spreading in the bottom part of the wall is limited. This predominance makes remarkable the difference between the perpendicular wall

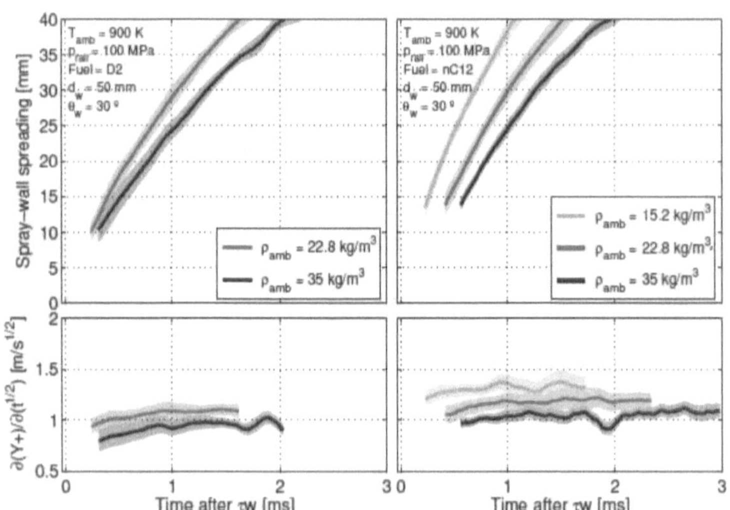

Fig. 5.12: Spreading along the wall (top) and its *R-parameter* (bottom) for different ambient densities and fuels ($T_{amb} = 900\,\text{K}$; $p_{rail} = 100\,\text{MPa}$; $d_w = 50\,\text{mm}$; $\theta_w = 30°$). Left: Spray of diesel #2. Right: Results with n-dodecane

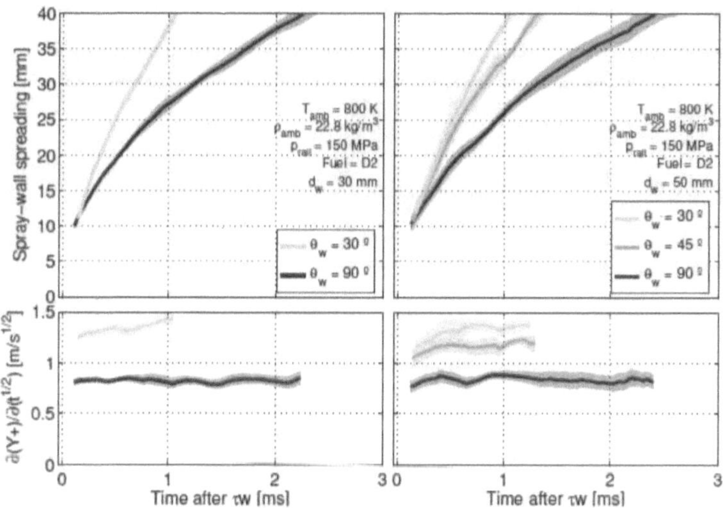

Fig. 5.13: Spreading along the wall (top) and its *R-parameter* (bottom) for different wall positions (T_{amb} = 800 K; ρ_{amb} = 22.8 kg m^{-3}; p_{rail} = 150 MPa; Fuel = D2). Left: Wall at 30 mm from the injector tip. Right: Wall positioned at d_w = 50 mm

spreading respect to the inclined wall one. In a shorter extent, even for inclined wall cases, the positive spreading is larger for narrower θ_w. In regards to the distance between the injector and the wall, it has negligible effect in the momentum (visible in *R-parameter*) of the spray for a certain direction and wall angle. This could be inferred from the conservation of momentum flux, considering that the momentum distribution along the wall happens in the same way for a given wall angle. This momentum deviation is a more relevant factor than the energy losses produced by the impact, taking into account that the pre-impinging velocity is higher for the wall at 30 mm. Considering this velocity prior reaching the wall, it highlights the fact of spray spreading along the wall being also similar with different 'initial conditions' that counteract with each other. Whereas in the 30 mm configuration, the spray reaches the wall with a higher velocity, the free spray at τ_w is wider for d_w = 50 mm, which gives to spreading a larger 'initial value' when collision occurs. However, it has to be taken into account that this similarity at different wall distances is observed with a temporal after-SWI-start phase and τ_w is strongly influenced by wall distance, this means that, as it could be predicted, since the beginning of an injection event, the spray will spread before onto a closer wall, but just as a result of the time gap between impacts and not because of differences in the spray general behavior.

Introducing another spray macroscopic feature, Figures 5.14, 5.15 and 5.16 depict, for different conditions, spray thickness measured at 10, 20 and 30 mm from the 'collision point', respectively referred to as Z_{th10}, Z_{th20} and Z_{th30}. In these plots, it can be appreciated how the general behavior of the curves is to start presenting a bump after the spray reaches the measuring point, due to the spray front vortex, and then to slightly stabilize in a nearly constant value. For these plots, another variation that can be seen is how thickness depends on the measuring point, presenting a thicker spray at larger distances and during the opening of the spray as it spreads along the wall. Particularly, Figure 5.14 shows thickness for gas temperature (left) and injection pressure (right) variations. Similarly as in the case of spreading, the gas temperature seems to have a negligible influence on the spray thickness, independently of how far it is measured. Nevertheless, thicknesses measured near from the 'impact point', tends to behave in a similar way when p_{rail} is changed, but if the measurement is made further, the lower the injection pressure is, the longer thickness takes to stabilize after the bump. On contrary, the higher the injection pressure is, the quicker the spray stabilization is reached, despite being more turbulent, and the spray is more 'pushed' against the wall reducing its thickness and increasing its spreading. Maximum thicknesses are quite similar independently of the injection pressure.

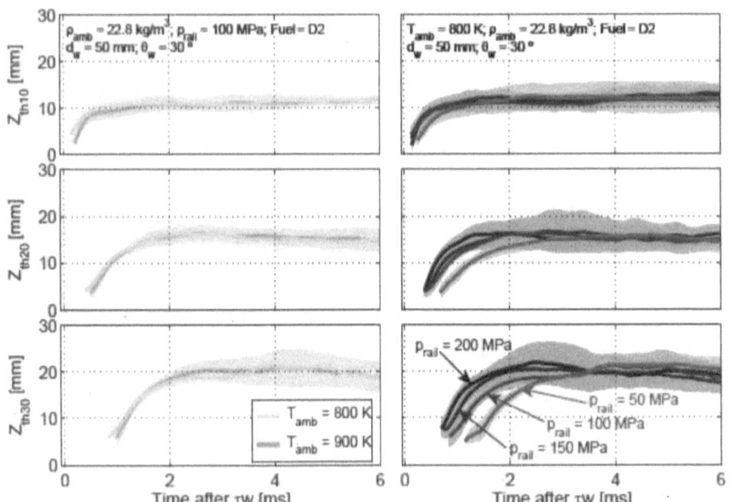

Fig. 5.14: Set of spray thicknesses measured at different wall points varying ambient temperature (left) and injection pressure (right) (ρ_{amb} = 22.8 kg m^{-3}; d_w = 50 mm; θ_w = 30°; Fuel = D2).

Density and fuel variations are shown in Figure 5.15. It can be seen, as it happens to the spray angle in free-jet conditions in a slighter extent, that density seems to promote a widening in spray thickness as a result of the enhancement in gas entrainment into the spray. Furthermore, the thickness bump duration is longer at higher densities for the same reason and also because the spray is going slower through the measuring point. Between different fuels however, there are no big differences that show to significantly go beyond the error fringe. Nonetheless, the tendency seen in Figure 5.15-right of the nC12 spray being thicker than the diesel one is consistent in most of points and it is in agreement with the behavior of spray angle observed in Figure 5.7, considering the different density ratio of the fuels. Given all this, the steady behaviors of thickness and spray angle with these parametric changes are quite analogous but are not as strong for the first of them, because of the partial exposition to ambient gas.

Regarding changes on the wall position, their effect on thickness of inert impinging sprays is shown in Figure 5.16. Stabilization in thickness is reached faster and at higher values at larger distances d_w. This could be explained by two different factors that apply for the nearest wall respect to the other: the higher velocity of collision of the spray, which makes the impact and spreading processes more turbulent and initially more unstable, and the shorter time

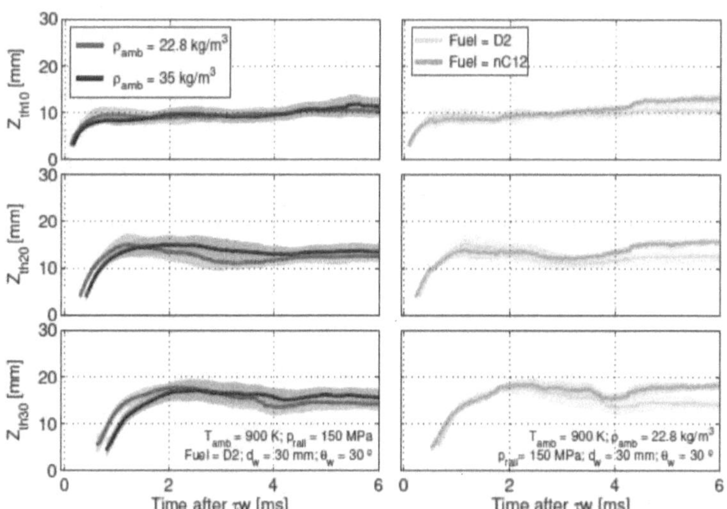

Fig. 5.15: Spray thicknesses for different gas densities (left) and fuels (right) at different distances from the 'collision point' ($T_{amb} = 900\,\mathrm{K}$; $p_{rail} = 150\,\mathrm{MPa}$; $d_w = 30\,\mathrm{mm}$; $\theta_w = 30°$).

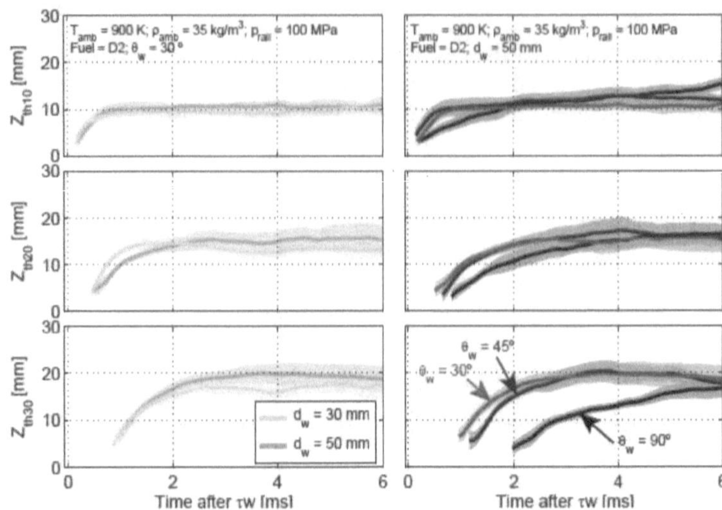

Fig. 5.16: Spray thickness varying wall position. ($T_{amb} = 900\,\mathrm{K}$; $\rho_{amb} = 35\,\mathrm{kg\,m^{-3}}$; $p_{rail} = 100\,\mathrm{MPa}$; Fuel = D2). Left: Different wall distances from the injector tip. Right: Variation of wall angle respect to nozzle hole axis.

the spray has gas entrainment under free-jet conditions. Right set of plots in Figure 5.16 depicts the effect of varying the wall angle, showing how the steady value of the thickness is quite similar for all of them, but it is reached differently by each one. The spray that interacts with a wall at $\theta_w = 90°$ has a slow stabilization while the inclined cases get steady significantly prompter. The same is observed in a shorter extent between the 30° and the 45° cases, where the first one stabilizes faster. This behavior is explained by the less dramatic deviation of the flow, due to the lesser disturbances in the spray during its interaction with the wall. The greater the wall angle is, the more transitory the behavior of the spray-wall interaction is at its beginning.

5.3 Free liquid spray visualization

This section is aimed to analyze the liquid spray interaction with the wall in terms of spray break-up and fuel evaporation, and how they are affected by altering operating and wall geometrical parameters of the tests. This is made through the visualization of the liquid phase of the spray by the use of the DBI technique, which allows to obtain images and contours as the shown in the samples of Figure 5.17. Before getting into SWI situations, this study required the prior understanding of the liquid free-jet. Figure 5.18 top set of plots shows the liquid penetration behavior vs. time, without a wall at different conditions. As predictable, a higher ambient temperature dramatically accelerates liquid evaporation rate as a consequence of the intensification of the kinetic energy of molecules at the spray surface, which shortens the distance that the liquid core reaches. Injection pressure, on the other side, does not have a considerable effect on liquid length. As observed in literature [21–23], the fuel injection rate rises at higher rail pressures while the gas entrainment increases in an equivalent way, maintaining the energy balance and therefore the liquid length. In the right-top plot, it can be seen how density increases gas entrainment and aerodynamically promotes jet break-up, having a strong role in the reduction of liquid length. From the fuel comparison side, their volatility is the main driver in getting different liquid lengths. N-dodecane has a lower overall distillation temperature than diesel, and therefore, a higher tendency to vaporize and to present a shorter liquid length. A last relevant consideration from the time resolved curves is that liquid penetration shows, always during the steady injection interval, a slight growth until reaching a constant value. Several works [7, 24] have found a similar behavior as a product of fuel temperature being higher at the injection start, due to the injector tip direct contact with the high temperature gases while, during the injection, the fuel flows through the injector and cools the tip down, reducing nozzle and fuel average temperatures and finally, increasing the liquid length.

This most stable and representative value of liquid length is found to be between 2 and 4 ms. Therefore, this interval is employed to obtain a mean

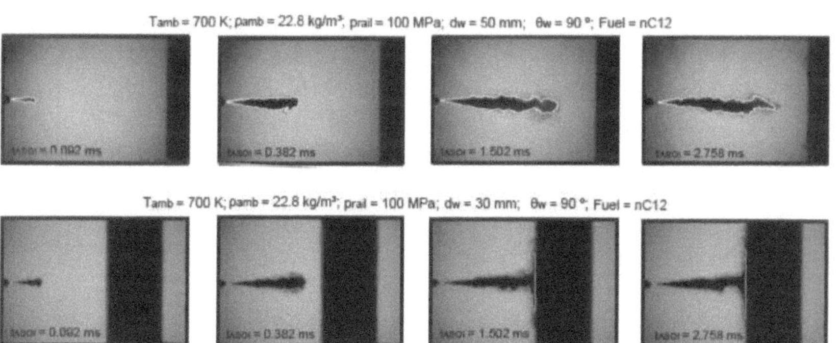

Fig. 5.17: Samples of random reps of the images observed via DBI (T_{amb} = 700 K; ρ_{amb} = 22.8 kg m^{-3}; p_{rail} = 100 MPa; θ_w = 90°; Fuel = nC12). Top set: Wall at d_w = 50 mm. Bottom set: Wall at 30 mm from the injector tip.

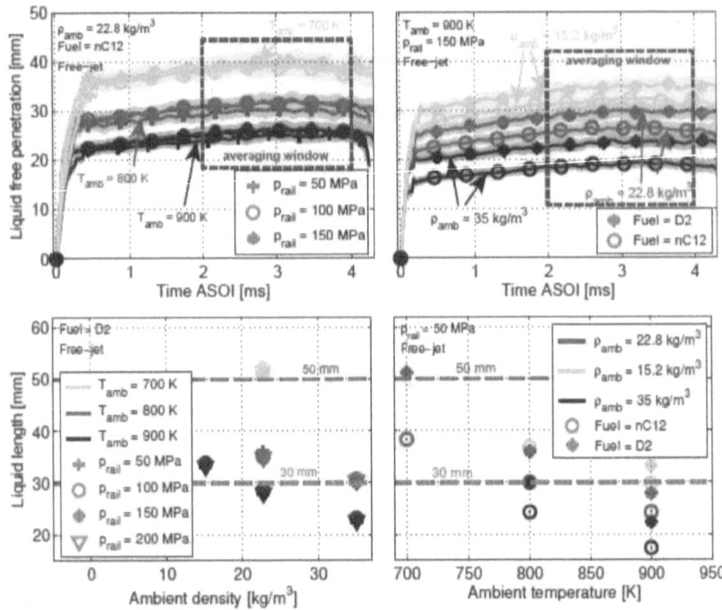

Fig. 5.18: Liquid length at free-jet conditions. Top set: Time resolved liquid penetration at different T_{amb} and p_{rail} (left) and varying fuels and ρ_{amb} (right). Bottom set: Averaged liquid length at a fixed fuel (left) and a single injection pressure (right).

liquid length *LL*, which is represented in the Figure 5.18 bottom set. The trends aforementioned are kept, but now it can be seen how the effect on *LL* that is produced by varying ambient temperature and fuel are significantly weaker at short liquid lengths. Broadly-speaking it can be said that, while liquid length is an energy-balance-driven characteristic, vapor spray evolution is driven by spray momentum. Dashed lines are located at 30 and 50 mm from the injector tip to identify the injector-wall distances present in the test plan. As seen, the entire text matrix barely have points that reach the 50 mm wall and the largest liquid lengths are seen at 700 K, a temperature that was not matched with all possible combinations. Therefore, in this case the adequate set of points suitable for the study of liquid spray-wall interaction is more limited than for vapor phase analysis, and the 30 mm wall distance will be the most valuable to extract information.

5.4 Liquid spray-wall interaction

Figure 5.19 shows a comparison of the contours detected through Schlieren imaging and DBI for vapor and liquid spray phases respectively, for a 'cold' evaporative ambient condition (both contours belong to the same random repetition). In the leftmost graph, evaporation is still not limiting the spray advancement, whose macroscopic structure is mainly liquid and the two contours are similar. In the second plot, where liquid steadiness has not been reached yet, there is a remarkable difference between vapor and liquid contours: liquid spreading is shorter than for a vapor spray, and the cloud of vapor after impingement is pretty much thicker than the liquid film. The third pair of contours has been taken after reaching steady state and the structure of liquid does not present much variation from the previous one: short spreading along the wall that reaches a certain value (analogously to liquid length in free-jet situations), a thin film in the order of 2 mm, and a stable spray angle upstream from the wall that is not different from the vapor one. The effect of varying the wall configuration on liquid spray spreading is depicted in Figure 5.20, where additionally the effect of injection pressure is shown. Only two injection pressures have been included to avoid an excess of redundant curves, as both the behavior and stable spreading reached by the liquid spray are similar for all injection pressures. This fact throws an interesting conclusion: Not only the null effect of injection pressure on liquid spreading is analogous to the found on free-jet liquid length because of the presence of the same evaporative mechanisms, but the secondary break-up generated by the jet-wall impact does not cause shortening on liquid steady spreading by itself, regardless of the spray velocity at the collision instant. The turbulence and energy dissipated in the impact act similarly as injection

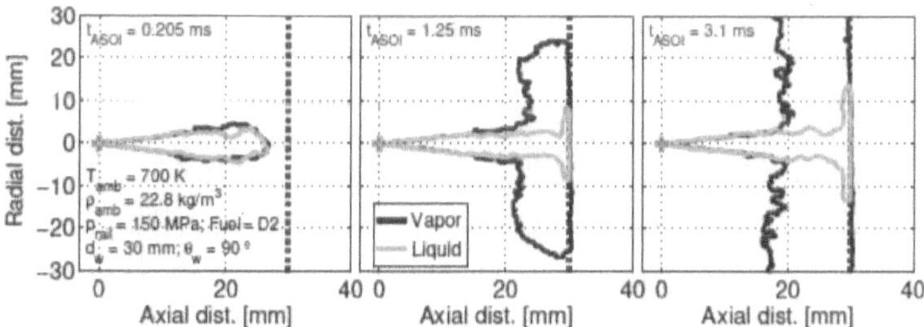

Fig. 5.19: Contours of a colliding inert spray comparing both liquid and vapor phases for the same condition (T_{amb} = 700 K; ρ_{amb} = 22.8 kg m^{-3}; p_{rail} = 150 MPa; Fuel = D2; d_w = 30 mm; θ_w = 90°).

pressure in the nozzle outlet, while in cold non-evaporative conditions they could affect liquid length reached by a jet, here there is a huge amount of evaporative energy available in the gas-fuel interaction and mixing. Both gas entrainment and fuel injection rate, rise and both maintain the balance of energy and therefore liquid length remains unchanged at free-jet conditions [6, 25–27]. At SWI conditions, the balance between the droplets break-up given by the impact and the fuel injection rate is also maintained and liquid spreading is essentially evaporation-driven [28, 29]. Studies on stable liquid spreading in SWI regimes are not easily found in literature. However, a study performed by [28] found similar liquid spreadings at p_{rail} = 50 MPa and 100 MPa using commercial diesel.

From the side of wall positioning effects, Figure 5.20-left shows the evident differences on liquid spreading from varying nozzle-wall distance, as in the 30 mm case, the jet widely impacts on the wall and has a wide spreading onto it, while in the 50 mm case, the liquid phase of the spray barely enters in contact with the wall. The right plot shows how, as it could be expected from the results obtained for the vapor phase, the liquid spray has a faster spreading and a larger steady length in the preferable direction in which it finds the least opposition respect to its original direction, which is defined by the inclination of the wall.

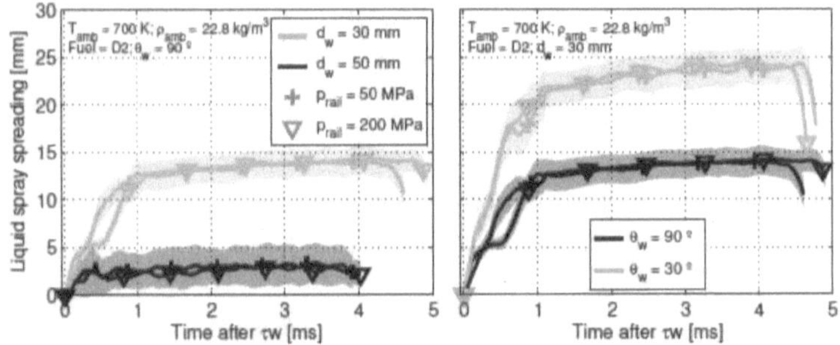

Fig. 5.20: Liquid spray spreading for different wall configurations and injection pressures (T_{amb} = 700 K; ρ_{amb} = 22.8 kg m^{-3}; Fuel = D2). Left: Wall distances comparison. Right: Different wall angles.

In Figure 5.19 the profile of a liquid spray is shown in comparison with a vapor contour. As has been previously mentioned, the spray contour seen through Schlieren visualization is not a transversal cut of the spray, as its internal thickness is covered by the vapor cloud formed by the spray front that goes in all directions. Nevertheless, that does not happen in the case of liquid sprays: the 'trumpet shaped' macroscopic structure of liquid spray has not important elements that could cover the central plane of the spray. Given this, and under the assumption of an axisymmetric spray when $\theta_w = 90°$, an estimation of the liquid spray volume can be made by considering it as the volume of the solid obtained by revolving π radians both the bottom and top halves of the detected jet contour. Liquid spray volume is an interesting metric to provide a robust analysis essentially for two reasons: The elusive analysis of points with a very short spreading (barely entering in contact with the wall as the 50 mm in Figure 5.20-left), and whose standard deviation is a large percentage of its mean value; and the definition of spray volume itself, which can be calculated for free-jet conditions too and allows a direct comparison between this situation and spray-wall interaction. Figure 5.21 shows the liquid spray volume of different 'axisymmetric' sprays. As seen for liquid spreading, the injection pressure effect on spray volume is negligible, not finding a significant influence in the whole steady liquid jet. The evaporation regime of nC12 is found again to be more intense than the observed for diesel, which means a shorter steady volume but a larger transitory growth from the start of injection, as seen in vapor penetration and spreading analyses previously discussed. On the other side, points with different chamber temperatures and nozzle-wall distances are shown in Figure 5.21-right, including free-jet conditions. As it could be expected from liquid length results, liquid spray volume is strongly reduced by rising ambient temperature. Interestingly, while injection pressure does not affect steady liquid volume, wall distance has an appreciable effect on it. This means that the reason is not associated to a higher pre-impingement velocity of the spray, as it could be firstly inferred. A better explanation is derived from the difference of the spray surface that is directly exposed to the hot atmosphere. The spray-gas contact area grows

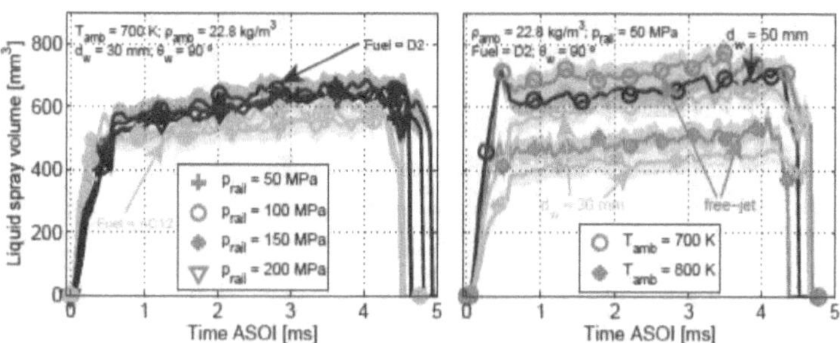

Fig. 5.21: Estimated liquid spray volume for axisymmetric sprays (free-jet and sprays against $\theta_w = 90°$ walls). Left: Comparison at different fuels and injection pressures. Right: Varying temperatures of the ambient and wall distance, including free-jet case.

when the spray collides in comparison of a free-jet straight plume, which promotes gas entrainment into the spray, enhancing heat transfer from the gas to the spray and increasing the evaporation rate. This rise on spray-gas contact area is given before and is more significant if the wall is closer to the nozzle outlet of the injector and, in any SWI case, it provokes a shorter liquid volume than a free-jet situation under the same injection and in-chamber conditions.

5.5 Modeling impinging spray behavior

An adequate prediction of the previously shown parameters is of great interest for several applications such as injection systems and/or combustion chambers design. Fast prediction models can also be helpful as preliminary estimations to refine more sophisticated simulation means, such as CFD. A numerous amount of laws and correlations have been proposed by researchers to predict the behavior of sprays at free-jet conditions. Naber and Siebers [12] proposed one of the most extensively exploited models in research for spray free penetration, based on the fundamental physics of gas-spray interaction by differencing two phases with different behaviors: a transitory zone with a linear proportionality to t_{ASOI} and a steady developed spray region with penetration proportional to $t_{ASOI}^{0.5}$. Several papers have found a similar behavior in this steady part, with models that include more parametrical variations and that are good agreement with experimental results [22, 30–32]. In some works that compound the present book [14, 33] the concept of *R-parameter* has been introduced from the well demonstrated proportionality of steady-spray penetration with time powered to 0.5. Siebers [25] defined a s caling law for liquid length based on mixing-limited vaporization. Higgins et al. [26] presented a different m odel t hat s uits b etter f or m ulti-component f uels. A considerable improvement of that correlation is given in [21]. Additionally, Payri has presented several interesting one-dimensional models to estimate liquid length [21, 24, 34]. Nevertheless, models for spray-wall interaction are less abundant in literature. Some works [29, 35] analyzed spray radial expansion along a wall in cold conditions based on parametrical variations. Some other models use a droplet-collision approach to propose correlations for a macroscopic spray development [36–38]. More sophisticated CFD approaches are more widely used in order to assess simulations at more realistic conditions [19, 39, 40].

Table 5.3 shows the standard form of the correlations proposed by the present book for each parameter and the different coefficients that distinguish

Table 5.3: Summary of the models created for the different parameters obtained from experimental results. Variables inputs and outputs are all in MKS system.

$$\text{PARAMETER} = \beta_0 \cdot \rho_{amb}^{\beta_1} \cdot \Delta p^{\beta_2} \cdot T_{amb}^{\beta_3} \cdot d_w^{\beta_4} \cdot sin^{\beta_5}(\theta_w) \cdot d_{Zth}^{\beta_6}$$

Coefficients	β_0	β_1	β_2	β_3	β_4	β_5	β_6	R^2
Input units	[–]	[kg m^{-3}]	[Pa]	[K]	[m]	[°]	[m]	[%]
***R-parameter* from spray free penetration $\partial(S_v)/\partial(t^{1/2})$ [m/s$^{1/2}$]**								
D2 (No-wall)	0.03858	-0.245	0.255	0	0	0	0	97.9
nC12 (No-wall)	0.04074	-0.245	0.255	0	0	0	0	98.4
D2 (Any-wall)	0.0390	-0.245	0.255	0	0	0	0	97.4
nC12 (Any-wall)	0.04127	-0.245	0.255	0	0	0	0	97.8
Liquid free length LL [m]								
D2 (No-wall)	26927	-0.4274	0	-1.82	0	0	0	96.4
nC12 (No-wall)	31528	-0.5436	0	-1.812	0	0	0	97.3
Spray vapor angle ϕ_v [°]								
D2 (No-wall)	7.9347	0.2948	0	0	0	0	0	96.1
nC12 (No-wall)	9.4367	0.2478	0	0	0	0	0	96.0
D2 (Any-wall)	8.1220	0.2948	0	0	0	0	0	89.5
nC12 (Any-wall)	9.6537	0.2478	0	0	0	0	0	92.2
Time of start of SWI (ASOI referenced) τ_w [s]								
D2	93.134	0.5025	-0.541	0	-0.125	0	0	99.1
nC12	92.924	0.5025	-0.541	0	-0.125	0	0	98.8
***R-parameter* from spray vapor spreading $\partial(Y_+)/\partial(t^{1/2})$ [m/s$^{1/2}$]**								
D2	0.01602	-0.245	0.255	0	0	-0.6082	0	91.2
nC12	0.01613	-0.245	0.255	0	0	-0.6082	0	89.8
Steady spray thickness along wall Z_{th} [m]								
D2	0.1605	0.16	0	0	0.388	-0.095	0.442	90.1
nC12	0.1675	0.16	0	0	0.388	-0.095	0.442	88.7

them, where Δp is the difference between injection pressure and back-pressure in the chamber, expressed in Pa and d_{Zth} is the measuring distance of spray thickness along the wall (10 mm; 20 mm and 30 mm as shown in Figures 5.14; 5.15 and 5.16) inputted in meters. The form of the formulas has been made generic for all parameters, in order to simplify the analysis and to easily compare the effect of each variation on the calculated parameter. This approach has been kept after ensuring that a good agreement between the measured and the modeled parameters is achieved. As previously shown, and as it is also visible in Table 5.3, spray angle is significantly affected by the density and fuel properties, similarly as found in [13, 18, 22], and the effect captured in the equation of the table has been implicitly included in the correlations of other parameters to keep their standard form, by not expressing them in function of $tan(\phi_v/2)$ as some of them (i.e. spray free penetration) are commonly represented [12, 31]. This also makes the use of the model faster to use because all parameters can be directly estimated without performing before the measurement or calculation of any other. Another consideration to be taken into account, is that variables that have not been altered in the whole test plan (such as injector geometrical features, wall material and/or quality related characteristics, fuel cooling settings, etc) are not expressed in the correlations and are implicitly included into the different coefficients. A similar approach has been followed with fuel properties, obtaining different coefficients for each one, depending on how much fuel affects each parameter. β_0 is the main coefficient used to absorb that kind of variations and also the differences between having a completely free-jet situation and the inclusion of a wall in any position when it is not important. The β_i coefficients that are equal to zero in Table 5.3, are considered not to have statistical significance (high p-value) and therefore are excluded from the correlation.

Variables with a significant effect on spray vapor penetration are both ambient density and injection pressure, as shown in Table 5.3. These coefficients and the obtained effects are in fair match with the obtained by several researchers like Naber and Siebers [12] and Desantes et al. [31]. The variation between the different correlations is given by the β_0 which is slightly different between fuels and even less variable if a wall is into the chamber, but still in accordance with the results shown in Figure 5.1. The liquid length correlation has a high coefficient of determination R^2 with slight variations in β_3, in contrast to β_0 and β_1 as both fuel density and fuel viscosity have an effect on LL but the strongest one is given by the evaporative properties of the fuel. While τ_w is penetration-like affected, the most significant effect seen in spray spreading is given by the angle of impingement, that deviates more or less the original momentum of the spray, and its thickness is considerably sensitive to the point of measure and to the distance between the wall and the nozzle outlet. Figure 5.22 serves as a summary of the regressions capability to predict the experimental values, showing a high confidence degree represented by the narrow area occupied by points around the $x = y$ line and the generally high correlation factor obtained for all models. It is important to highlight that these correlations have been constructed from a huge range of conditions covered by a total of 196 samples.

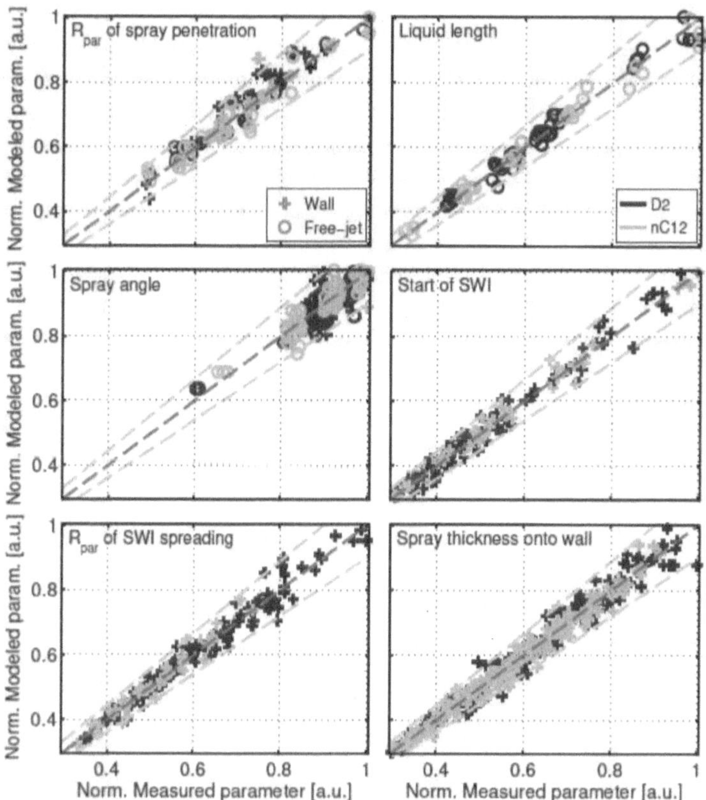

Fig. 5.22: Evaluation of the results from models developed for different variables versus its measured values. The dark gray dashed line represent model = experiment while the lighter lines cover the area of a 10% deviation.

5.6 Conclusions

The effect of operating conditions and wall features over the spray development and its interaction with a wall is experimentally studied under evaporative and non-reactive situations. Macroscopic spray characteristics were obtained by using independent cameras and different optical techniques to get high-speed simultaneous images of both liquid and vapor phases of the spray at diesel-like conditions, suppressing autoignition by carrying out the experiment with nitrogen as ambient gas. The results of this study are synthesized in this chapter.

While vapor penetration is highly affected by the momentum flux of the spray, whose balance is given by injection pressure and ambient density, liquid phase characteristics are driven by vaporization mechanisms that affect gas-jet heat transfer, such as gas density and mainly, its temperature. Regarding free-jet conditions, the spray angle is primarily controlled by ambient density due to the larger gas entrainment into the vapor spray. Nevertheless, the sole presence of the wall into the chamber has a slight but perceptible effect on upstream spray angle, in comparison with a no-wall arrangement: Spray angle widens with the opening of the spray after it impacts against the wall and opens itself due to its larger area exposed to the ambient gases respect to the free-jet. However, there are not significant impact on spray angle with the wall configuration of angle or distance.

Inert spray spreading along the wall can be considered analogous to spray free penetration, even remaining proportional to the square root of time in its steady stage, besides both being similarly affected by injection pressure and gas density. For inclined walls, the spreading presents a preferred direction that corresponds to the one that makes the widest angle between the original spray axis and the wall plane, and its velocity increases as the wall is more inclined, which lets the perpendicular case as the one with the slowest but most symmetrical spreading. Spray thickness was measured at different distances from the 'collision point' being similarly susceptible to parametric variations in comparison to the spray angle and it is appreciably affected by nozzle-wall distance but, even more, by the distance from the 'collision point' in which it is measured. In regards to liquid phase characteristics under SWI, injection pressure has a negligible influence, while the shorter the wall distance is, the less voluminous the steady state spray gets. Correlations for several parameters were developed and they seem to adequately predict the spray behavior observed in the experiments with high degree of confidence.

The findings presented in this chapter can be used for 1-D and CFD model validation in order to provide a solid base to understand the mechanisms that drive the observations of the spray-wall interaction at evaporative inert conditions.

References

[1] Giraldo, Jhoan S, Payri, Raul, Marti-Aldaravi, Pedro, and Montiel, Tomas. "Effect of high injection pressures and ambient gas properties over the macroscopic characteristics of the diesel spray on multi-hole

nozzles". In: *Atomization and Sprays* 28.12 (2019), pp. 1145–1160 (cited on pages 109, 112).

[2] Bardi, Michele et al. "Engine Combustion Network: Comparison of Spray Development, Vaporization, and Combustion in Different Combustion Vessels". In: *Atomization and Sprays* 22.10 (2012), pp. 807–842 (cited on pages 109, 112).

[3] Meingast, Ulrich, Staudt, Michael, Reichelt, Lars, and Renz, Ulrich. "Analysis of Spray / Wall Interaction Under Diesel Engine Conditions". In: *SAE Technical Paper 2000-01-0272* 724 (2000), pp. 1–15 (cited on page 109).

[4] Kook, Sanghoon and Pickett, Lyle M. "Liquid length and vapor penetration of conventional , Fischer-Tropsch , coal-derived , and surrogate fuel sprays at high-temperature and high-pressure ambient conditions". In: *Fuel* 93 (2012), pp. 539–548 (cited on page 109).

[5] Westlye, Fredrik R et al. "Penetration and combustion characterization of cavitating and non-cavitating fuel injectors under diesel engine conditions". In: *SAE Technical Paper 2016-01-0860* (2016), p. 15 (cited on page 109).

[6] Siebers, Dennis L. "Liquid-Phase Fuel Penetration in Diesel Sprays". In: *SAE Technical Paper 980809* (1998), pp. 1–23 (cited on pages 109, 128).

[7] Gimeno, Jaime, Martí-Aldaraví, Pedro, Carreres, Marcos, and Peraza, Jesús E. "Effect of the nozzle holder on injected fuel temperature for experimental test rigs and its influence on diesel sprays". In: *International Journal of Engine Research* 19.3 (2018), pp. 374–389 (cited on pages 109, 125).

[8] Payri, Raul, Gimeno, Jaime, Bracho, Gabriela, and Vaquerizo, Daniel. "Study of liquid and vapor phase behavior on Diesel sprays for heavy duty engine nozzles". In: *Applied Thermal Engineering* 107 (2016), pp. 365–378 (cited on page 109).

[9] Pickett, Lyle M, Genzale, Caroline L, Manin, Julien, Malbec, Louis-Marie, and Hermant, Laurent. "Measurement Uncertainty of Liquid Penetration in Evaporating Diesel Sprays". In: *ILASS Americas 23rd Annual Conference on Liquid Atomization and Spray Systems.* Ventura, CA (USA): ILASS-Americas, 2011 (cited on page 109).

[10] Som, Sibendu and Aggarwal, Suresh K. "Assessment of Atomization Models for Diesel Engine Simulations". In: *Atomization and Sprays* 19.9 (2009), pp. 885–903 (cited on page 109).

[11] Allocca, Luigi, Lazzaro, Maurizio, Meccariello, G., and Montanaro, Alessandro. "Schlieren visualization of a GDI spray impacting on a heated wall: Non-vaporizing and vaporizing evolutions". In: *Energy* 108 (2016), pp. 93–98 (cited on page 109).

[12] Naber, Jeffrey D and Siebers, Dennis L. "Effects of Gas Density and Vaporization on Penetration and Dispersion of Diesel Sprays". In: *SAE Paper 960034* (1996) (cited on pages 109, 112, 114, 130, 132).

[13] Agarwal, Avinash Kumar and Chaudhury, Vipul H. "Spray characteristics of biodiesel/blends in a high pressure constant volume spray chamber". In: *Experimental Thermal and Fluid Science* 42 (2012), pp. 212–218 (cited on pages 109, 132).

[14] Gimeno, Jaime, Bracho, Gabriela, Martí-Aldaraví, Pedro, and Peraza, Jesús E. "Experimental study of the injection conditions influence over n-dodecane and diesel sprays with two ECN single-hole nozzles. Part I: Inert atmosphere". In: *Energy Conversion and Management* 126 (2016), pp. 1146–1156 (cited on pages 109, 112, 115, 130).

[15] ECN. *Engine Combustion Network*. Online. 2010 (cited on page 109).

[16] Database., PubChem. *National Center for Biotechnology Information. Dodecane, CID=8182.* Online (cited on page 109).

[17] Martí-Aldaraví, Pedro. "Development of a computational model for a simultaneous simulation of internal flow and spray break-up of the Diesel injection process". PhD book. Valencia: Universitat Politècnica de València, 2014 (cited on page 112).

[18] Payri, Raul, Salvador, Francisco Javier, Gimeno, Jaime, and Viera, Juan Pablo. "Experimental analysis on the influence of nozzle geometry over the dispersion of liquid n-dodecane sprays". In: *Frontiers in Mechanical Engineering* 1 (2015), pp. 1–10 (cited on pages 112, 132).

[19] Yu, H. et al. "Numerical Investigation of the Effect of Alcohol-Diesel Blending Fuels on the Spray-Wall Impingement Process". In: *SAE Technical Papers* 2016-April.April (2016) (cited on pages 114, 130).

[20] Delacourt, E, Desmet, B, and Besson, B. "Characterisation of very high pressure Diesel sprays using digital imaging techniques". In: *Fuel* 84.7-8 (2005), pp. 859–867 (cited on page 114).

[21] Payri, Raul, Viera, Juan Pablo, Gopalakrishnan, Venkatesh, and Szymkowicz, Patrick G. "The effect of nozzle geometry over the evaporative spray formation for three different fuels". In: *Fuel* 188 (2017), pp. 645–660 (cited on pages 125, 130).

[22] Xu, Min, Cui, Yi, and Deng, Kangyao. "One-dimensional model on liquid-phase fuel penetration in diesel sprays". In: *Journal of the Energy Institute* 89.1 (2016), pp. 138–149 (cited on pages 125, 130, 132).

[23] Pickett, Lyle M, Genzale, Caroline L, and Manin, Julien. "Uncertainty quantification for liquid penetration of evaporating sprays at diesel-like conditions". In: *Atomization and Sprays* 25.5 (2015), pp. 425–452 (cited on page 125).

[24] Payri, Raul, Garcia-Oliver, Jose Maria, Bardi, Michele, and Manin, Julien. "Fuel temperature influence on diesel sprays in inert and reacting conditions". In: *Applied Thermal Engineering* 35 (2012), pp. 185–195 (cited on pages 125, 130).

[25] Siebers, Dennis L. "Scaling liquid-phase fuel penetration in diesel sprays based on mixing-limited vaporization". In: *SAE Technical Paper 1999-01-0528* (1999) (cited on pages 128, 130).

[26] Higgins, Brian S, Mueller, Charles J, and Siebers, Dennis L. "Measurements of Fuel Effects on Liquid-Phase Penetration in DI Sprays". In: *SAE Technical Paper 1999-01-0519* (1999) (cited on pages 128, 130).

[27] Viera, Juan Pablo. "Experimental Study of the Effect of Nozzle Geometry on the Performance of Direct-Injection Diesel Sprays for Three Different Fuels". PhD book. Universitat Politècnica de València, 2017 (cited on page 128).

[28] Zhang, Yanzhi et al. "Numerical and experimental study of spray impingement and liquid film separation during the spray/wall interaction at expanding corners". In: *International Journal of Multiphase Flow* 107.June (2018), pp. 67–81 (cited on page 128).

[29] Trujillo, M. F., Mathews, W. S., Lee, C. F., and Peters, J. E. "Modelling and experiment of impingement and atomization of a liquid spray on a wall". In: *International Journal of Engine Research* 1.1 (2000), pp. 87–105 (cited on pages 128, 130).

[30] Araneo, Lucio, Coghe, Aldo, Brunello, G, and Cossali, Gianpietro E. "Experimental Investigation of gas density effects on diesel spray penetration and entrainment". In: *SAE Paper 1999-01-0525* (1999) (cited on page 130).

[31] Desantes, Jose Maria, Payri, Raul, Salvador, Francisco Javier, and Gil, Antonio. "Development and validation of a theoretical model for diesel spray penetration". In: *Fuel* 85.7-8 (2006), pp. 910–917 (cited on pages 130, 132).

[32] Desantes, Jose Maria, Payri, Raul, Salvador, Francisco Javier, and Gimeno, Jaime. "Prediction of Spray Penetration by Means of Spray Momentum Flux". In: *SAE Technical Paper 2006-01-1387* (2006) (cited on page 130).

[33] Payri, Raul, Gimeno, Jaime, Peraza, Jesús E., and Bazyn, Tim. "Spray / wall interaction analysis on an ECN single-hole injector at diesel-like conditions through Schlieren visualization". In: *Proc. 28th ILASS-Europe, Valencia* September (2017) (cited on page 130).

[34] Payri, Raul, Salvador, Francisco Javier, Gimeno, Jaime, and Zapata, Luis Daniel. "Diesel nozzle geometry influence on spray liquid-phase fuel penetration in evaporative conditions". In: *Fuel* 87.7 (2008), pp. 1165–1176 (cited on page 130).

[35] González, Uriel. "Efecto del choque de pared en las características del chorro Diesel de inyección directa". PhD book. Valencia: E.T.S. Ingenieros Industriales. Universitat Politècnica de València, 1998 (cited on page 130).

[36] Gavaises, Manolis, Theodorakakos, Andreas, and Bergeles, George. "Modeling wall impaction of diesel sprays". In: *International Journal of Heat and Fluid Flow* 17.2 (1996), pp. 130–138 (cited on page 130).

[37] Stanton, Donald W. and Rutland, Christopher J. "Multi-dimensional modeling of thin liquid films and spray-wall interactions resulting from impinging sprays". In: *International Journal of Heat and Mass Transfer* 41.20 (1998), pp. 3037–3054 (cited on page 130).

[38] Andreassi, L, Ubertini, S, and Allocca, Luigi. "Experimental and numerical analysis of high pressure diesel spray-wall interaction". In: *International Journal of Multiphase Flow* 33.7 (2007), pp. 742–765 (cited on page 130).

[39] Zhang, Yanzhi et al. *Development of anew spray/wall interaction model for diesel spray under PCCI-engine relevant conditions.* Vol. 24. 1. 2014, pp. 41–80 (cited on page 130).

[40] Zhao, L. et al. "An Experimental and Numerical Study of Diesel Spray Impingement on a Flat Plate". In: *SAE Int. J. Fuels Lubr.* 10.2 (2017), pp. 407–422 (cited on page 130).

Chapter 6

Reactive spray against an ambient-temperature wall

The results of the second experimental campaign in the HPHTV are gathered and condensed in this chapter. In this occasion, the vessel is filled with flowing standard air extracted from the atmosphere, then xO_2 is approximately 0.21. Because of that, injection and spray-wall interaction take place in a reacting environment that promotes ignition occurrence. In spite of having the presence of combustion in these experiments, the quartz plate remains as the wall to be in interaction with the spray. Due to the prolonged time in which in-chamber conditions are established before measuring a point, the exposure of all the faces of the wall to the hot ambient and the lack of a heat sink (i.e. a fluid flowing along any wall face) [1–3] that could cool the wall down, the plate is considered to be at the same temperature than the ambient ($T_w \approx T_{amb}$). This fact makes this experiment a half way between an inert impinging spray and a reactive one against a wall with engine-like temperature and heat transfer behavior [4–7].

Boundary conditions are similar to the presented in the previous chapter, as can be seen in Table 6.1. The same two fuels n-dodecane and diesel #2 are used for this campaign, this time with a typical dry air oxygen concentration in the ambient. The target gas densities were maintained respect to the previous experiment by adjusting the ambient pressure for each temperature and the consideration of standard air. The wall is set at the same positions, adding a $d_w = 50$ mm - $\theta_w = 60°$ configuration, in order to make a better link between this chapter and the configurations available for the next one, since the QT-Wall hardware is more versatile.

Table 6.1: Test plan for reactive tests in the quartz wall.

Variable	Values	Units
Wall hardware	None (free-jet) - QT-Wall	-
Fuel	nC12$^{\otimes}$ - D2	-
Injector	Bosch 3-22 ECN Spray D (D103)	-
Energizing time (ET)	2.5	ms
Tip temperature (T_{tip})	363	K
Oxygen fraction (xO_2)	≈0.21	-
Gas temperature (T_{amb})	800 - 900	K
Gas density (ρ_{amb})	22.8 - 35	kg/m^3
Injection pressure (p_{rail})	50 - 100 - 150 - 200$^{\triangledown}$	MPa
Wall distance (d_w)	30 - 50	mm
Wall angle (θ_w)	30 - 45$^{\triangle}$ - 60$^{\triangle}$ - 90	°
Total test points	216	points

$^{\otimes}$ Not all possible combinations with other variables were tested
$^{\triangledown}$ Only for diesel (D2) tests.
$^{\triangle}$ Only for $d_w = 50$ mm. Please refer to Figure 3.5-right

Despite the test matrix has been narrowed with the exclusion of low-temperature and the low-density conditions respect to the non-reacting campaign, the amount of extracted data is considerably larger due to the use of a more complex optical setup, capable of the simultaneous employment of four cameras instead of two, as shown in Figure 4.5: Two cameras are used to record the flame natural luminosity at high-speed from the side and front through

Table 6.2: Details of the optical setup by technique (Figure 4.5).

Feature	Schlieren imaging	N.L. (side)	N.L. (frontal)	OH* chemilumin.
Camera	Photron SA-X2	Photron SA5	Photron SA5	Andor-iStar
Sensor type	CMOS	CMOS	CMOS	ICCD
Filter CWL	480 nm	390 nm	390 nm	310 ± 5 nm
Shutter time	3.28 μs	19.25 μs	19.25 μs	-
Frame rate	40000 fps	25000 fps	25000 fps	1 frame/inj.
TTL-Width	-	-	-	2.5 ms$^{\triangledown}$
Pixels/mm ratio	5.88	7.00	9.60	8.82
Reps per point	8	8	8	8

$^{\triangledown}$The TTL-Delay was set from 1.5 to 3 ms (ASOE) depending on the ignition delay.

the transparent wall; another fast camera acquired images of the vapor phase of the spray from the side by the use of the Schlieren technique as done in the previous chapter, and finally, an intensified ICCD camera was used to observe the flame lift-off length via OH* chemiluminescence. Table 6.2 provides details about the optical setup where the ICCD camera was configured in a range dependent on the delay of the ignition, in order to always trigger acquisition before combustion start with similar advance between points (TTL-Delay is respect to SOE). Note that the ICCD camera was slightly inclined (about 6°) to avoid blocking the field of view of the others. Nevertheless, this angle is small and the possible effects were taken into account by properly correcting the acquired images.

6.1 Ignition delay

The ignition delay is the time from the start of the injection to the high-temperature reactions that lead to the combustion of the fuel-air mixture. In the literature, many ways to characterize ignition delay can be found [8–11]. Due to the high-intensity luminosity (as a product of soot) of diesel flames, a processing method based on intensity measurement inside the spray area was used as mentioned in chapter 4, considering the start of the second stage of ignition as ignition delay or *ID*.

Figure 6.1 shows ignition delay variation with the different controlled variables at different fixed conditions. Specifically, the left set of plots shows how ignition delay is affected by operating conditions when a determined fuel is injected at a fixed wall position. Similarly as literature indicates for free-jet cases, ignition delay is strongly shortened by ambient temperature, as could be expected from the higher rate of evaporation which enhances the mixture process as seen in the previous chapter. In parallel, the interaction between fuel and air at higher densities promotes a better mixing process and a larger oxygen availability. A quicker liquid break-up is given due to the larger turbulences at higher injection pressures, resulting into a decrease of ignition delay. Both effects of density and injection pressure are less significant at short *ID* cases, where the high reactivity produced by the high temperature is the main factor due to the acceleration of the reactions of oxidation. Both cases shown are in the presence of the quartz wall, and qualitatively, there are not big differences on the trends that could be expected from free-jet.

The right set of Figure 6.1 shows the dependence of ignition delay on wall conditions and fuel properties, including the no-wall case. The high volatility of dodecane respect to diesel #2, shown in the previous chapter

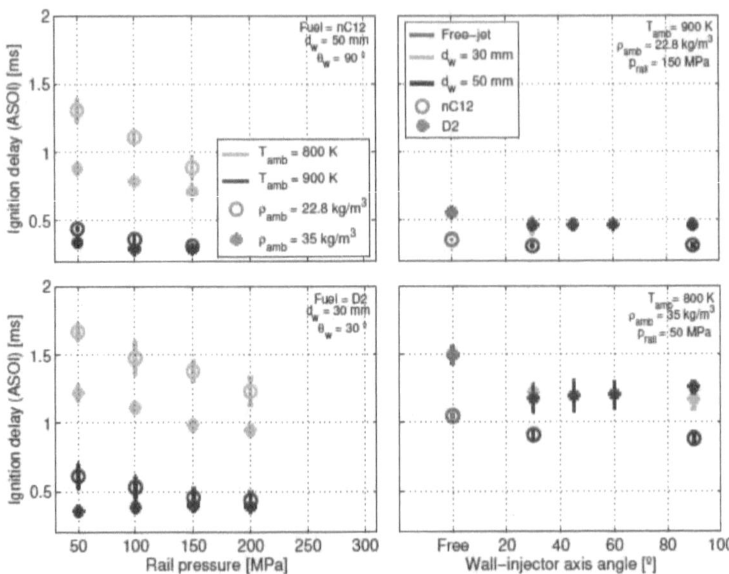

Fig. 6.1: Ignition delay calculated for different conditions. Left set: *ID* vs injection pressure varying ambient temperature and density. Right set: Variation of ignition delay vs. wall angle at different wall orientations and fuels (free-jet included).

in terms of liquid phase behavior, makes it prone to ignite before. This observation is in accordance with the results and with the cetane indexes shown in Table 3.2, being nC12 cetane number approximately two times the diesel one. Wall angle does not seem to have a significant effect on ignition delay. By previously observing that the narrower the wall angle, the larger the stable liquid spreading, it could lead to think that ignition delay should diminish with angle widening. However, rather than an improvement in the evaporation process due to a better atomization, this happens solely by the jet deviation on the wall that, when it is inclined, has a preferential direction while in a perpendicular wall the liquid spreads in all directions. An adequate further analysis to confirm it could be to compute the liquid volume of a spray that impinges onto an inclined wall, as done for the $\theta_w = 90°$ wall, by the use of a different optical technique. With regard to the droplet break-up caused by the wall, it does not present representative changes with collision angle, but its presence definitively makes a difference respect to the free-jet case, where the lack of this secondary break-up induced by the beginning of SWI delays the ignition.

This statement is better illustrated in Figure 6.2, where the observed ignition delay in all impinging spray tests is plotted vs. the same conditions in free-jet cases. Short ignition delays are quite similar and are affected in the same extent by the fuel and injection conditions, since they take place before the spray reaches the wall, not having spray-wall interaction even when there is a wall and a collision occurring after *ID*. As ignition delay becomes longer, it can be seen that it starts to be affected by the wall, due to the acceleration in the reactions produced by the mixing improvement, exhibiting an *ID* up to the 85% respect to the free-jet case. Additionally, ignition delay seems to be consistently shorter for $d_w = 30$ mm. Although there is barely difference on liquid spreading with wall-injector distance, this is a stable value. On the other hand, the liquid jet impacts and its consequent droplet size reduction takes place before in the closest wall case, promoting a prompter ignition occurrence.

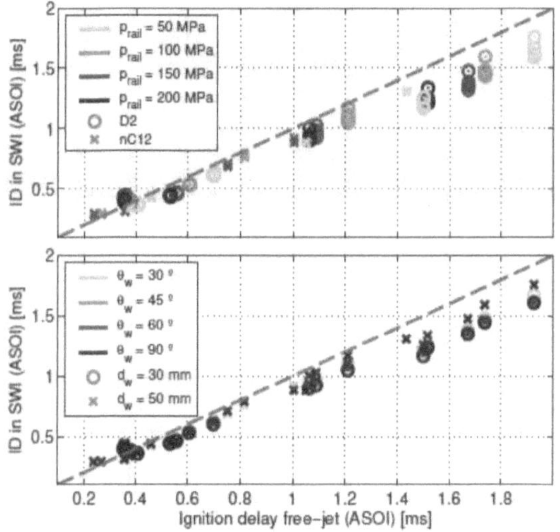

Fig. 6.2: Ignition delay with the quartz wall in SWI conditions vs. ignition delay at free-jet conditions. The gray dashed line represents $ID_{free\text{-}jet} = ID_{SWI}$. Both plots have the same information differently classified. Top: Fuel and injection pressure variation. Bottom: Changes in wall position.

6.2 Reacting free spray visualization via Schlieren

The study of the spray morphology of the gaseous phase of the spray is based, like in the previous chapter, on the Schlieren visualization of the injection and the spray-wall interaction. An example of the images and the contour obtained for the geometrical analysis is given in Figure 6.3. In the first image it can be seen a pure vapor spray before ignition and afterwards, both the start of cool flames and the second phase of ignition taking place (dark spray degradation and a following expansion with a visible light release). The use of this optical approach implies that the contour covers both the vapor phase of the fuel and the gaseous burned products of combustion that are similarly part of the spray. The behavior of the contour observed via Schlieren and covered by this definition, is what is analyzed along the chapter and is what hereafter is referred to as 'vapor' to be in accordance with the observations of the non-reacting campaign, even thought the encompassed concept is slightly wider.

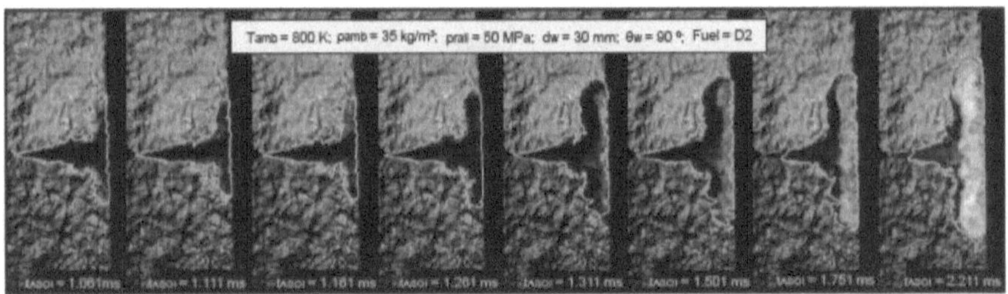

Fig. 6.3: Series of spray images and contours of a random sample of the reacting spray recorded via Schlieren (T_{amb} = 800 K; ρ_{amb} = 35 kg m^{-3}; p_{rail} = 50 MPa; d_w = 30 mm; θ_w = 90°; Fuel = D2).

In the following graphs, ignition delay is shown together with the curve of its respective parameter (represented with a circle), and with its standard deviation in bars. This has been done in order to trace the start of ignition to follow the process of spray development and spray-wall interaction. The lack of an *ID* mark is an indication of an ignition delay not occurring in the time range shown in the plot. Spray penetration before reaching the wall is compared at both inert and reacting conditions in Figure 6.4. In the experimental work done in [12], where a free-jet study has been conducted using the same injector, it is explained in detail that a reactive spray does not present a constant *R-parameter* but it has a behavior which can be divided in the following phases: a constant zone before ignition, where it behaves as an inert spray, a bump due to the expansion that is produced when ignition takes place, a valley where *R-parameter* drops, and finally, a quasi-steady phase with a stable value that is higher than the inert one. This behavior, although limited due to the end of free-jet evolution regime when the wall is reached, can be observed in

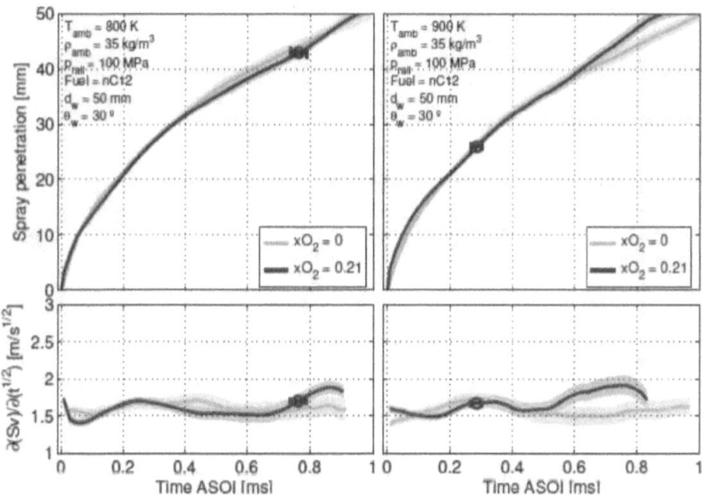

Fig. 6.4: Free penetrations (top) and their *R-parameter* (bottom) for both reacting and inert conditions (different oxygen concentrations) and gas temperatures ($\rho_{amb} = 35\,\mathrm{kg\,m^{-3}}$; $p_{rail} = 100\,\mathrm{MPa}$; $d_w = 50\,\mathrm{mm}$; $\theta_w = 30°$; Fuel = nC12). Left: Test conditions at $T_{amb} = 800\,\mathrm{K}$. Right: Gas temperature $T_{amb} = 900\,\mathrm{K}$

the 900 K case, where ignition still occurs far upstream from the wall and the produced spray acceleration is more noticeable. Therefore, the spray reaches the wall (50 mm) before the 800 K case does. This obviously means that penetration is affected, not only by the same parameters as inert sprays, but it is also sensitive to variables that control ignition such as temperature [13].

This indirect effect of temperature on spray penetration by affecting ignition timing can also be observed in Figure 6.5. Injection pressure not only causes a faster penetration due to a higher spray momentum as seen in the inert experiments, but the combustion-driven boost produced in penetration occurs before at high p_{rail} for its ignition delay shortening. Figure 6.6 shows penetration with different fuels and ambient densities. For reactive cases, the most significant difference between sprays of different fuels is produced by the delayed ignition of diesel respect to dodecane. However, fuel properties still do not affect spray penetration as much as ambient density does in terms of spray-air momentum transfer. Free penetration is just shown for 50 mm, since it is the longest wall-injector distance. Within this limited range, ignition delays are relatively short and they are not dramatically different, making the effects of the sole ignition delay differences not to seem as strong as the aspects already known from the inert spray experience. However, the

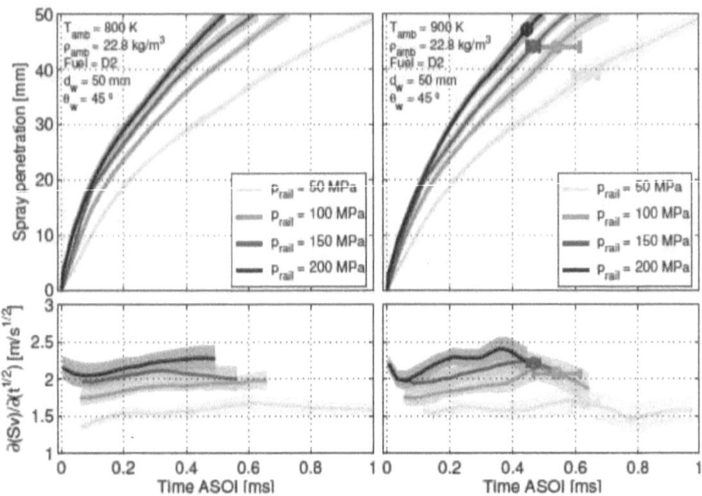

Fig. 6.5: Reactive free penetrations (top) and their *R-parameter* (bottom) for different injection pressures and ambient temperatures ($\rho_{amb} = 22.8\,\mathrm{kg\,m^{-3}}$; $d_w = 50\,{*}\mathrm{mm}$; $\theta_w = 45°$; Fuel = D2). Left: Air temperature at 800 K. Right: Vessel set at $T_{amb} = 900\,\mathrm{K}$

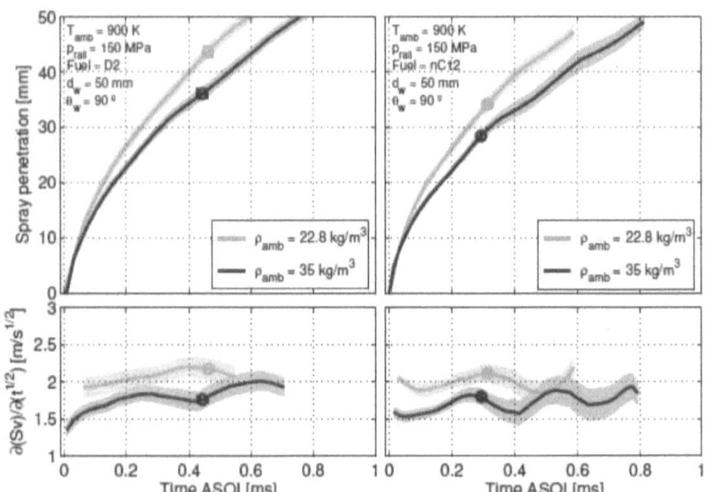

Fig. 6.6: Reactive free penetrations (top) and their *R-parameter* (bottom) for various fuels and air densities ($T_{amb} = 900\,\mathrm{K}$; $p_{rail} = 150\,\mathrm{mm}$; $d_w = 50\,\mathrm{mm}$; $\theta_w = 90°$). Left: Tests with diesel. Right: Points with dodecane as fuel

conditions of spray velocity and width in the very moment of impingement could be expected to be very different respect to non-reacting situations and SWI parameters could be affected by combustion occurrence [10, 13, 14].

6.3 Reacting spray-wall interaction via Schlieren

An analogous behavior is seen again between penetration and spray-wall spreading in Figure 6.7. Nonetheless, ignition delay taking place before reaching the wall or near of it, does affect the initial conditions of the spray-wall interaction. As seen in both cases, spreading already starts spatially advanced in reacting conditions respect to the inert ones, and it is affected by gas temperature because of ignition timing when is longer. This effect, as seen for spray penetration, acts in terms of the temporal gap between ignition delays and of its effect on the premixing time that enhances the following ignition-driven spray expansion. Therefore, regardless of the transitory phases of the spray, the quasi-steady reacting *R-parameter* is not dependent on ignition delay or τ_w in SWI situations.

In Figure 6.8, the effect of rail pressure can be appreciated. As previously discussed, ignition delay is shortened by injection pressure, however, τ_w is

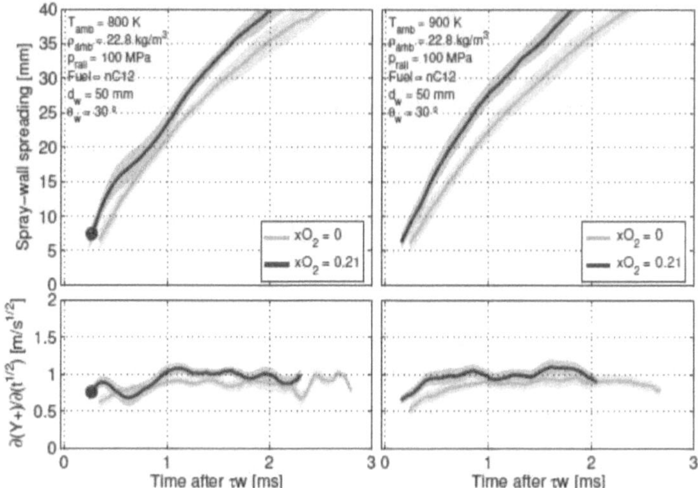

Fig. 6.7: Spray spreading along the wall (top) an its *R-parameter* (bottom) for both reacting and inert conditions (different oxygen concentrations) and gas temperatures (ρ_{amb} $= 22.8\,\mathrm{kg\,m^{-3}}$; $p_{rail} = 100\,\mathrm{MPa}$; $d_w = 50\,\mathrm{mm}$; $\theta_w = 30°$; Fuel = nC12). Left: Test conditions at $T_{amb} = 800\,\mathrm{K}$. Right: Gas temperature $T_{amb} = 900\,\mathrm{K}$

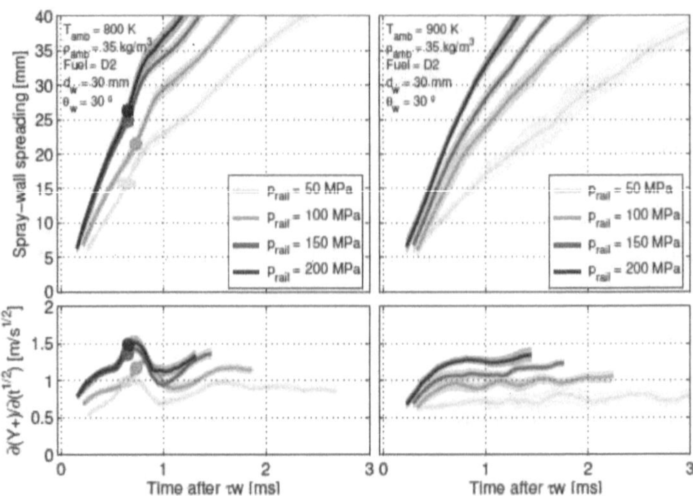

Fig. 6.8: Reacting spray spreading (top) and its respective *R-parameter* (bottom) for different injection pressures and ambient temperatures (ρ_{amb} = 35 kg m^{-3}; d_w = 30 mm; θ_w = 30°; Fuel = D2). Left: Air temperature at 800 K. Right: Vessel set at T_{amb} = 900 K

so. Therefore, in this particular case, ignition delays and *R-parameter* bumps seem to be phased in this temporal reference. Spray momentum as expected is proportional to *R-parameter* and the trends observed for inert sprays are kept in reactive conditions. On the other hand, Figure 6.9 shows the effect of changing fuels and gas densities. The faster advancement of the spray in low density environments is observed despite the increment on ignition delay seen in this type of atmospheres, and this trend is maintained in quasi-steady stages. However, the effect of fuel viscosity on spray spreading seen in inert ambients is negligible in comparison to the produced by the differences on ignition delay in terms of fuel reactivity, having a slight influence in stable conditions.

Figure 6.10 shows spray-wall spreading at different wall angles and with both left and right set of plots having different wall distances. It can be seen how the curves at θ_w = 90° start with an apparent initial value while inclined walls show a smoother beginning. This is due to the already existent spray width, which is totally projected as spreading quicky after collision time for a perpendicular impingement, while the projection of this width is lesser when the wall is inclined. This effect is more noticeable in reacting sprays whose width has been expanded before reaching the wall. Additionally, the effect of the inclination on the momentum distribution is observed to favor spray

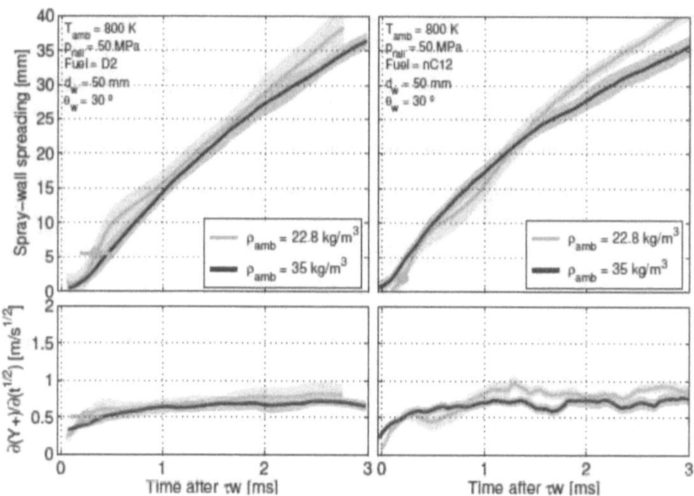

Fig. 6.9: Reacting spray spreading (top) and its respective *R-parameter* (bottom) for different air density and fuel (T_{amb} = 800 K; p_{rail} = 50 MPa; d_w = 50 mm; θ_w = 30°). Left: Points with Fuel = D2. Right: Points using n-dodecane as fuel.

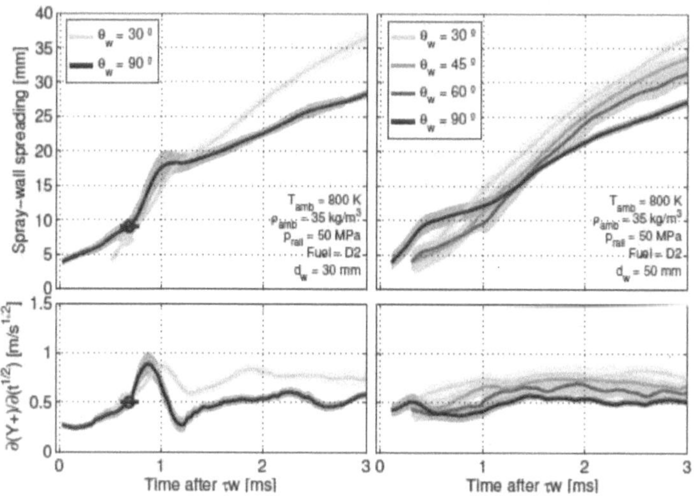

Fig. 6.10: Reactive wall spreading (top) and their *R-parameter* (bottom) for different wall positions (T_{amb} = 800 K; ρ_{amb} = 35 kg m^{-3}; p_{rail} = 50 MPa; Fuel = D2). Left: Tests with injector-wall distance of 30 mm. Right: Wall located at 50 mm from the injector tip.

spreading in the measuring direction. The only effect of wall distance on spreading is given by the relative ignition delay (or advance) respect to the start of SWI. In a $d_w = 50\,\mathrm{mm}$ the ignition delay is prone to occur before the spray hits the wall or at least to be less spread onto the wall.

Now, in regards to the spray thickness, Figure 6.11 makes a direct comparison between the behavior seen in the previous chapter for inert sprays and for sprays in a reacting atmosphere with an oxygen concentration of 20.9%. Both sets of plots are differenced in terms of long *ID* (low gas density and temperature) and short ignition delay (high-temperature and high-density ambient) to observe differences between tests with ignition taking place onto the wall or before it. As seen, spray thickness is the same for both atmospheres until ignition occurs. Once this happens, a strong combustion-induced expansion is observed, and it is progressively larger as the measuring distance is further from the 'collision point', which makes the increment of steady thickness with the measuring distance even more significant. On the other side, in the reacting test point of the right panel of graphs, combustion starts before the spray touches the wall, not only having a slight advancement in the $xO_2 = 0.21$ point to reach the wall and the

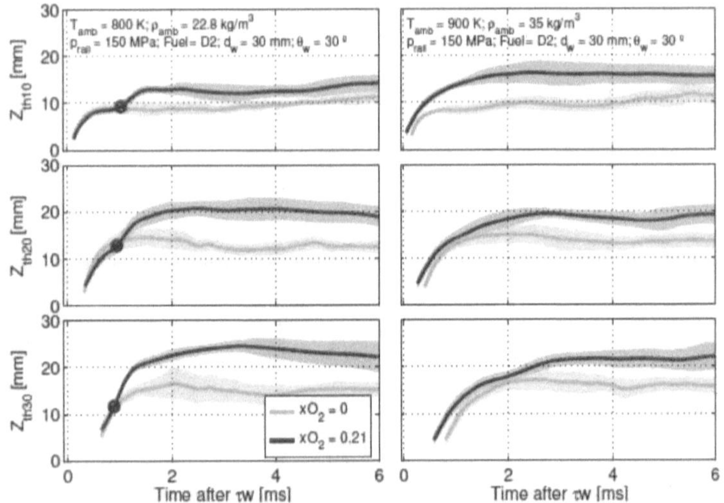

Fig. 6.11: Spray thickness varying oxygen concentration between inert and reacting conditions ($p_{rail} = 150\,\mathrm{MPa}$; $d_w = 30\,\mathrm{mm}$; $\theta_w = 30°$; Fuel = D2). Left: Long ignition delay conditions of gas temperature and density. Right: High temperature and density conditions (short *ID*).

measuring points, but also a thicker spray, which means that the effect of the ignition-driven expansion on thickness remains even when ignition starts before SWI and its consequent spray momentum deviation.

Ignition is more delayed at lower temperatures at isolating the effect of ambient temperature. Therefore, the fuel-air mixing before combustion is longer and the spray expansion shows to be stronger. Figure 6.12-left illustrates this, where steady thickness is larger for the low temperature case. Now, the influence of rail pressure and turbulence on ignition moment has been already stated to be solid but not as strong as the gas temperature or density ones. Given that, injection pressure shows a negligible influence on stable thickness in SWI. It can be observed that the thickness profile takes more to stabilize in reactive cases than in inert sprays due to the prolongation of the expansion, and the continuous spreading of combustion-produced gases from the reaction zone.

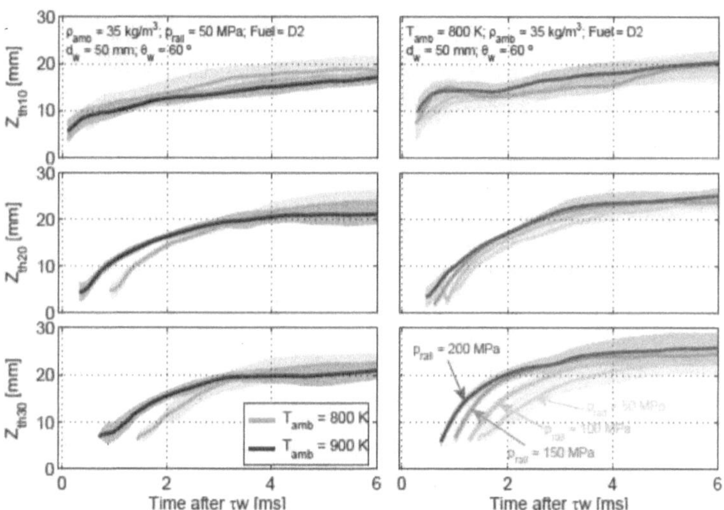

Fig. 6.12: Spray thickness for different ambient temperatures and injection pressures (ρ_{amb} = 35 kg m^{-3}; d_w = 50 mm; θ_w = 60°; Fuel = D2). Left: Temperature variation at rail pressure of 50 MPa. Right: Different injection pressures at a fixed gas temperature of 800 K.

Those gaseous products are still advancing within the chamber and are mixing with the surrounding air, whose entrainment into the mixture and the consequent spray thickness are incremented at higher densities, as depicted in Figure 6.13-left. This effect is stronger than for inert sprays and it is predominant over the aforementioned effect of ignition delay on thickness because of the lower density ratio ρ_f/ρ_{amb} of the products respect to the unburned fuel [15]. Thickness observed for nC12 and D2 in non-reacting conditions showed to be barely narrower for diesel. In Figure 6.13-right it can be observed how both effects of the reduction of density ratio after lift-off length and ignition timing gaps make differences between fuels to be more relevant in terms of their spray thicknesses.

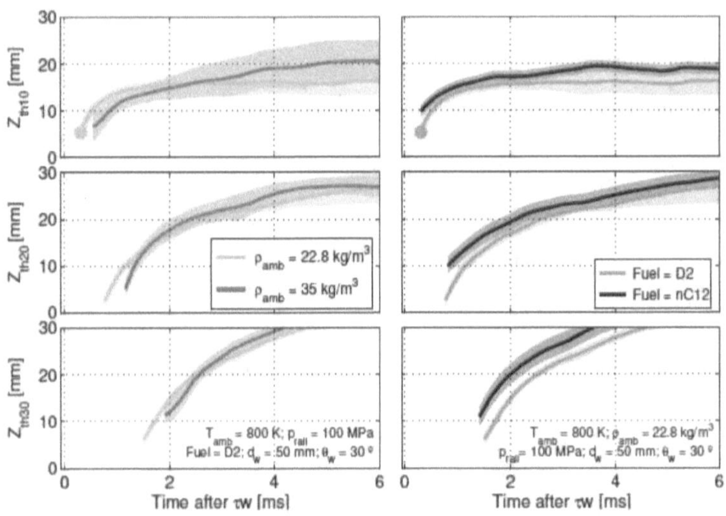

Fig. 6.13: Spray thickness for different gas densities and fuels. (T_{amb} = 800 K; p_{rail} = 100 MPa; d_w = 50 mm; θ_w = 30°). Left: Different gas densities using diesel as fuel. Right:ρ_{amb} = 22.8 kg m^{-3} and the two fuels.

The influence of wall position is condensed into Figure 6.14 where it was taken into account that ignition delays that are shorter than τ_w remain the same and then, they are not affected by the wall. Therefore, several points whose ignition starts before reaching the wall were selected for this set of graphs. The furthest wall case has a longer time after SOI to τ_w with air entrainment into the spray, the one that after combustion is composed by burned products with a lower density than the unburned fuel before ignition. This makes the characteristic larger thickness of reactive sprays at large wall distances to be more significant, as appreciable in Figure 6.14 left set of plots, even when the curves are in phase due to the similar behavior of spreading. Despite being incremented by the expansion induced by combustion, the same trends observed in inert spray remains with a slow thickness stabilization in perpendicular wall cases as the biggest difference with wall angle. It has to be taken into account that the observed thicknesses, specially in the 90° case, are given by the very front vortex that covers the spray in its periphery, which makes its stabilization to be apparently slower.

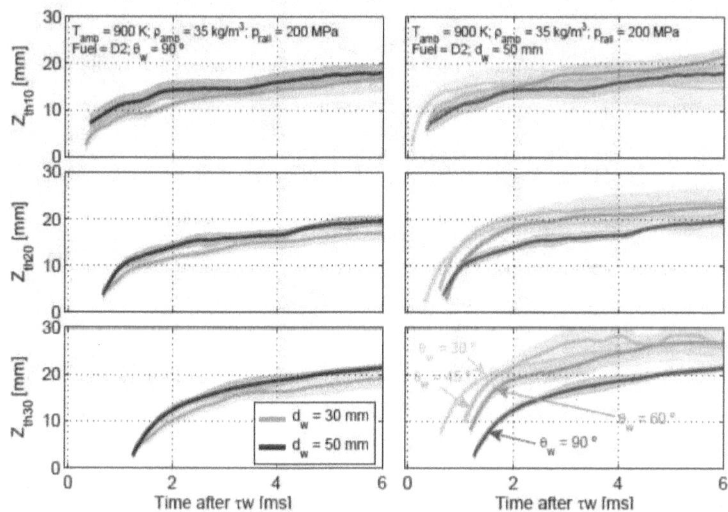

Fig. 6.14: Spray thickness for different wall positions. (T_{amb} = 900 K; ρ_{amb} = 35 kg m^{-3}; p_{rail} = 200 MPa; Fuel = D2). Left: Different wall distances from the injector tip for θ_w = 90°. Right: Different wall angles fixing d_w = 50 mm.

6.4 Hot soot flame direct visualization

Combustion flame luminosity was recorded in the same line of sight of Schlieren imaging, only with the use of a broadband filter to avoid saturation. A high shutter time was used to have a better visualization of the flame structure evolution as sampled with a raw image in Figure 4.13. It is important to highlight that the scope of this study was to characterize the effect of parametrical variations on the flame morphology in terms of its shape evolution. From that, no qualitative or quantitative analysis of incandescence intensity, optical length or soot concentration was carried out. Figure 6.15 shows another sample of different flame images obtained via natural luminosity and its respective detected contour. The images of a hot flame are taken for the same conditions varying only the inclination angle (from top to bottom) and progressively in time (from left to right), observing how the flame evolution is

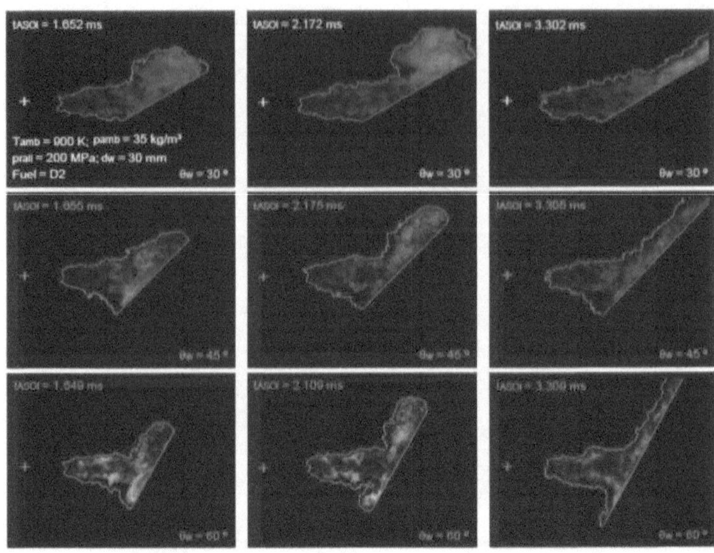

Fig. 6.15: Series of flame images observed via natural luminosity from the side view (T_{amb} = 900 K; ρ_{amb} = 35 kg m^{-3}; p_{rail} = 200 MPa; d_w = 50 mm; Fuel = D2). Top set: Wall inclination at 30°. Center set: Wall at θ_w = 45°. Bottom set: Wall inclined at θ_w = 60°.

similar as the observed one with previous techniques as Schlieren, at least in terms of going advancing towards a preferred direction due to wall inclination that favors spray velocity. However, it can be seen how the front vortex of the spray is significantly larger in height than the posterior thickness of the flame observed at long times from SOI. The following analysis tries to cover the study of all those flame characteristics.

It is known that at free-jet conditions the flame moves forward penetrating along with the vapor spray [16–18]. On the other hand, Figure 6.16 shows the flame spreading onto the wall along with vapor spreading observed via Schlieren imaging (the last one in dashed lines and with no representation of error fringes for data visualization purposes) for ignition taking place after (left plot) and before (right plot) SWI start. Again, what is referred to as 'vapor' strictly makes reference to the gaseous fuel observed via Schlieren imaging and includes burned combustion products. What is seen is not only that the same behavior of faster spreading at higher injection pressures remains, but also that a more sooty flame produced in an ambient at high temperature does not have to penetrate faster. Instead of that, as mentioned for free penetration, steady flame spreading goes together with the vapor front of the spray, whose parametrical variations have been discussed. In the left plot, that includes the ignition start event for the four test points, it can be seen how no flame is visible at the beginning of vapor spreading. Then, when a first set of pixels rises in intensity due to the exothermic reactions start, the natural luminosity curves begin at a point of the wall that depends on the location of these first illuminated pixels, to quickly expands along the wall to reach vapor spreading value. Another remarkable fact is that exposure time is high enough to allow detection of flame areas before the reported ignition delay, or that is to say, to detect light emission of chemical reactions between the 'start of cool flames' and the 'second stage of ignition' (SoCF and SSI respectively).

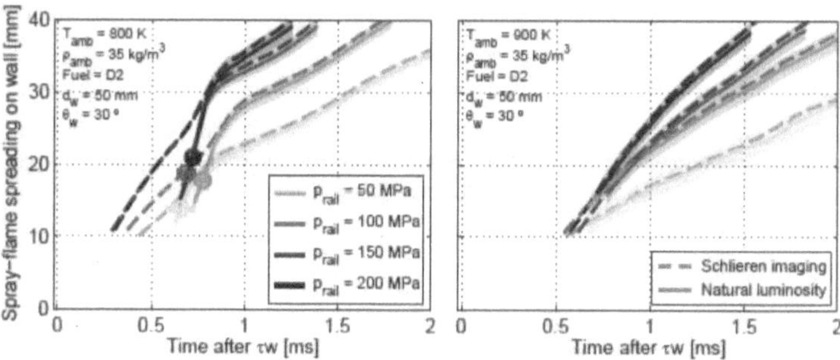

Fig. 6.16: Comparison of flame and vapor (Schlieren imaging) spray spreadings at different p_{rail} and T_{amb} ($\rho_{amb} = 35\,\text{kg m}^{-3}$; $d_w = 50\,\text{mm}$; $\theta_w = 30°$; Fuel = D2). Left: Gas temperature at 800 K. Right: $T_{amb} = 900\,\text{K}$. The shady error fringe is for the flame seen via natural luminosity.

Figure 6.17 helps to verify the agreement of steady phase spreading measured for both vapor and flame, showing the following expression in the vertical axis, which represents the maximum difference between vapor and flame spreadings found in the quasi-steady measured interval:

$$\Delta Y_{vap\text{-}fla} = max\left(\frac{Y_{+vap} - Y_{+fla}}{Y_{+vap}}\right) \cdot 100\% \qquad (6.1)$$

Where Y_{+vap} and Y_{+fla} are both respective spreadings. As seen there, absolute difference is below 2.5%, which is around 1.1 mm depending on the Y_+ value. These plots show that this similarity between spreadings is maintained also at different wall configurations, gas densities and fuel properties. This is not only reasonable and in agreement with a free-jet penetration expected behavior, considering a high shutter time, but it is also satisfactory due to the shown match between both visualizations in terms of experimental execution, processing procedure and general configuration selection. This low $\Delta Y_{vap\text{-}fla}$ difference is broadly below standard deviation of both parameters.

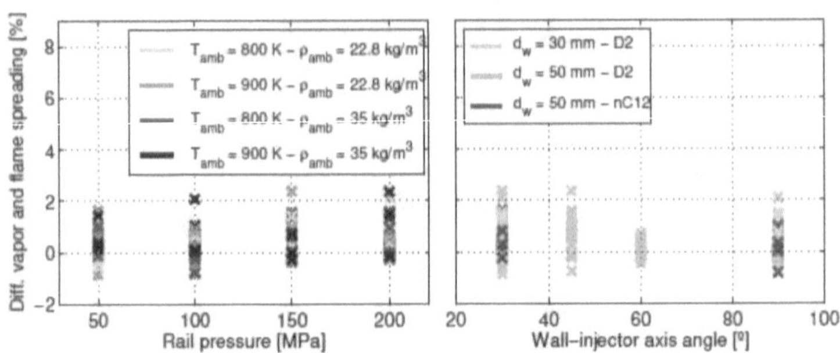

Fig. 6.17: Difference between spreading calculated for vapor (Schlieren) and sooting flame (natural luminosity) for different points. Left: Varying injection pressure, gas density and temperature. Right: variation of wall position and fuel.

Moreover, the flame spreading was observed from a frontal view through the transparent wall. Figures 6.18; 6.19 and 6.20 show the behavior of both horizontal spreadings towards the left and the right sides of the captured image (X_{f-} and X_{f+} respectively) in the top plots, and the flame width ΔX_f in the bottom ones. In the top plots, time (referenced after SWI start) has been located in the vertical axis, while horizontal flame spreading has been placed in the horizontal one. Therefore, flame expansion in the two horizontal orientations is seen to advance towards the plot sides, centering the graph at the 'collision point' and taking leftwards as negative and rightwards as positive. Bottom plots axes orientation is standard, with time in the X-axis and flame-wall width in the Y-axis.

Figure 6.18 shows this graph arrangement with a fixed density, fuel and wall angle and distance; varying gas temperature and rail pressure. The first observation to remark is that both leftwards and rightwards spreadings are quite symmetrical. From the known effects of gas temperature and wall angle, this symmetry indicates both a proper alignment of the wall and an adequate homogeneity of ambient conditions inside the vessel, taking into account that temperature characterization in the HPHTV was not made with thermocouples located along the X_f axis. These effects remain as known in horizontal spreading: high injection pressure accelerates the flame front spreading in all wall directions. High ambient temperature makes flames to appear sooner. Nevertheless, the flame growth of the more premixed spray at low gas temperature is more sudden and abrupt.

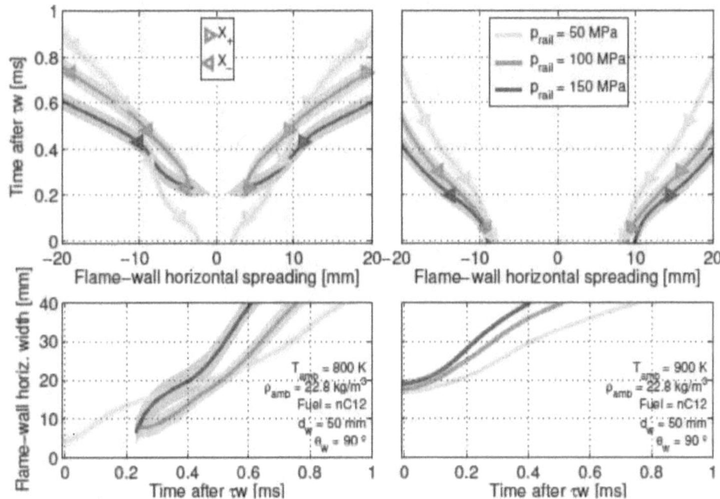

Fig. 6.18: Flame horizontal spreadings and width for different p_{rail} (ρ_{amb} = 22.8 kg m^{-3}; d_w = 50 mm; θ_w = 90°; Fuel = nC12). Left: T_{amb} = 800 K. Right: T_{amb} = 900 K.

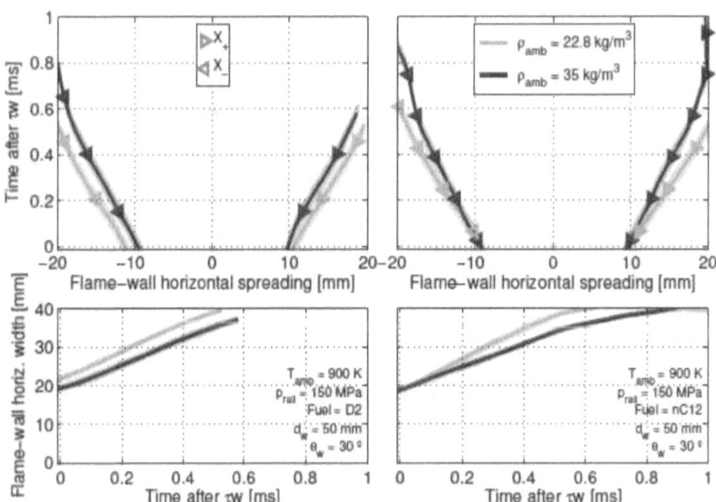

Fig. 6.19: Flame horizontal spreadings and width for different ρ_{amb} (T_{amb} = 900 K; p_{rail} = 150 MPa; d_w = 50 mm; θ_w = 30°). Left: Fuel = D2. Right: Fuel = nC12.

Horizontal spreading at different densities and fuels was measured, and it is depicted in Figure 6.19 while the effect of varying wall position is shown in Figure 6.20. Most of the expected effects from the similarities between sooty flame and vapor spreading development are observed. Nonetheless, wall inclination angle has no effect on this horizontal spreading due to the coincidence of the axis in which is measured and the rotation axis of the wall at different configurations: regardless of the wall inclination, the horizontal axis projection is always at 90° respect to the free spray axis. Despite visible horizontal range is limited to 40 mm and it is quickly reached in the fastest stages of SWI, it is still enough to appreciate how trends are visible before steady phase and consistent with the natural luminosity observed from the side view.

Time-resolved thickness of the sooty flame onto the wall was also determined at the same three distances 10 mm; 20 mm and 30 mm from the 'collision point'. Figure 6.21 shows points at different air temperatures (left) and injection pressures (right). Respect to the observed vapor thickness, flame is narrower and has a front vortex quite more prominent. High temperature promotes not only higher levels of soot concentration observable in terms of light intensity, but also in terms of thickness. Higher temperatures reduce air/fuel mixture time and therefore the diffusion combustion is increased

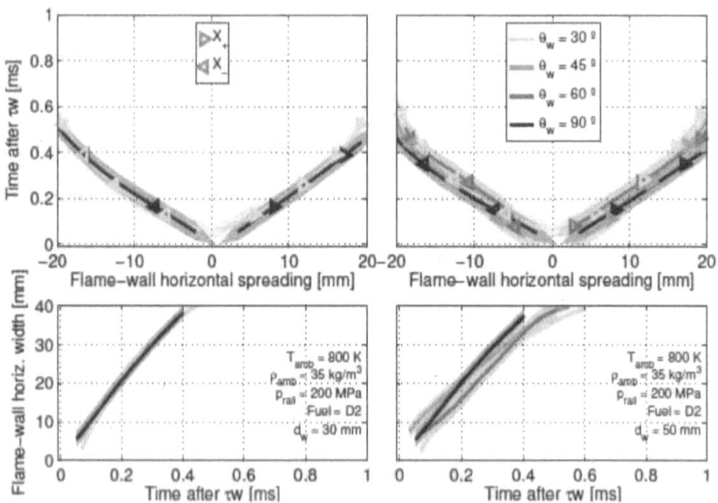

Fig. 6.20: Flame horizontal spreadings and width varying wall angle (T_{amb} = 800 K; ρ_{amb} = 35 kg m^{-3}; p_{rail} = 200 MPa; Fuel = D2). Left: d_w = 30 mm. Right: d_w = 50 mm.

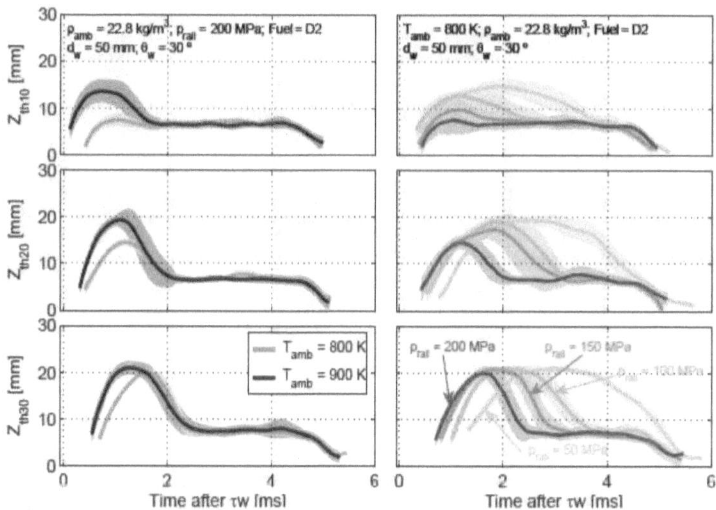

Fig. 6.21: Flame thickness for different ambient temperatures and injection pressures (ρ_{amb} = 22.8 kg m^{-3}; d_w = 50 mm; θ_w = 30°; Fuel = D2). Left: Temperature variation at rail pressure of 200 MPa. Right: Different injection pressures at a fixed gas temperature of 800 K.

respect to the premixed one. Additionally it promotes the burning of surrounding unburnt fuel. On the other side, rail pressure reduction conducts to a thicker flame along the wall, mainly as product of three factors: A slower flame spreading feeds up the flame vortex and moreover, it takes a longer time to go through the measuring point. Furthermore, high rail pressure during the injection event is 'pushed' against the wall and; finally, high injection pressures are advantageous in reducing soot formation due to atomization and turbulence enhancement. The differences between points on the vortex height are reduced as the flame spreads further along the wall.

This reduction on the vortex height differences indicates that, such as it generally happens for the steady part after the vortex, flame thickness is similar regardless of injection pressure and these differences are given from the transitoriness in the SWI start, and therefore it is reduced as the spray advances. The same trend is observed in Figure 6.22-left, where two points at different air densities are shown. Flame vortex is greater for high density conditions when is measured at 10 mm and its growth rate is reduced in comparison to the 22.8 kg m^{-3} case. Fuel variation is seen in Figure 6.22-right and there is not a significant difference on thicknesses either in the vortex

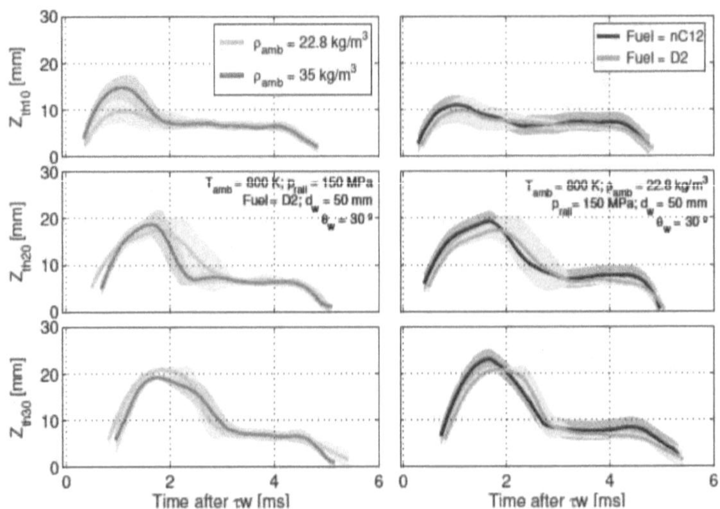

Fig. 6.22: Spray thickness for different gas densities and fuels. (T_{amb} = 800 K; p_{rail} = 150 MPa; d_w = 50 mm; θ_w = 30°). Left: Different gas densities using diesel as fuel. Right:ρ_{amb} = 22.8 kg m^{-3} and the two fuels.

nor the stable height. Although n-dodecane is a sooty fuel [12, 19], at similar operating conditions the flame shape evolution is not substantially different after ignition, but vortex duration is slightly shorter than in diesel, accordingly to spray tip velocity.

Wall position is varied in the tests shown in Figure 6.23. The larger the extent of premixed combustion and air entrainment surface are (in 'after τ_w reference', taking into account that the flame reaches before the wall at 30 mm than at 50 mm) the thicker is the flame and the larger its vortex is, as a consequence of the wall distance. On the other side, more inclined walls result into thicker flames in terms of steady and flame front vortex. In this case, flame distribution changes with wall angle, feeding up the flame growth with the higher spreading velocity and the less transitory behavior of the most inclined cases to the perpendicular one, where the flame is equally distributed onto the wall in all directions.

6.5 Lift-off length

As mentioned in chapter 4 and illustrated with Figure 4.19, the lift-off length of a flame that impinges on a wall at d_w = 30 mm is very susceptible to be

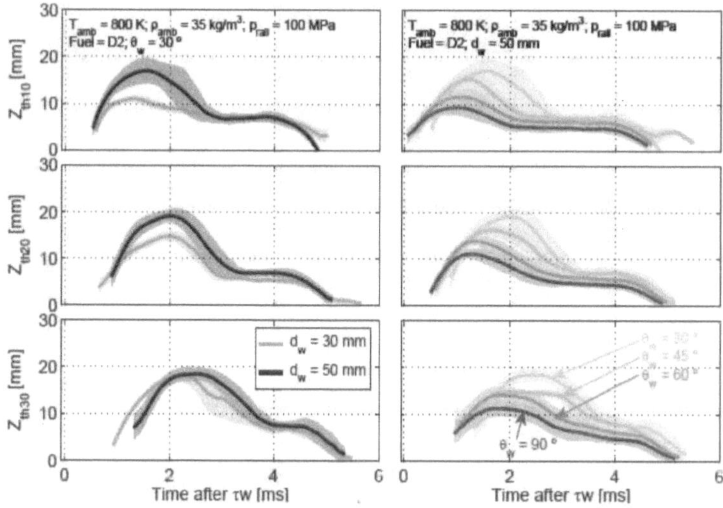

Fig. 6.23: Flame thickness for different wall positions. (T_{amb} = 800 K; ρ_{amb} = 35 kg m^{-3}; p_{rail} = 100 MPa; Fuel = D2). Left: Different wall distances from the injector tip for θ_w = 30°. Right: Different wall angles fixing d_w = 50 mm.

covered by the thickness of the flame observed via OH* chemiluminescence. Taking into account this large diameter injector hole, it can be presumed that even some lift-off length are long enough to be covered with a wall at 50 mm from the nozzle outlet, since this flame height is normally around 15-20 mm length. A first observation of the optically accessible *LoLs* is made in Figures 6.24 and 6.25. Several authors [8, 20–22] indicate that lift-off length is strongly controlled by ignition delay. From there, it can be understood that lift-off length is short for elevated high gas temperatures and densities. This same link between ignition delay and lift-off length explains that, for iso-density and iso-temperature ambient conditions and considering that the effect of p_{rail} on *ID* is not as strong as the others, at higher injection velocities (i.e. injection pressures), similar ignition delays are given when sprays are traveled longer distances and lift-off length are larger. Moreover, differences in fuel reactivity affect lift-off length too, which is always shorter for nC12 than for diesel as shown in the right set of plots.

The wall angle not only does not have apparent effect on lift-off length, but also those values shown in Figure 6.24 (d_w = 50 mm) present the same lift-off length than free-jet conditions. Unfortunately, almost all d_w = 30 mm points of the test matrix have their *LoL* covered by the very flame. Nevertheless, all

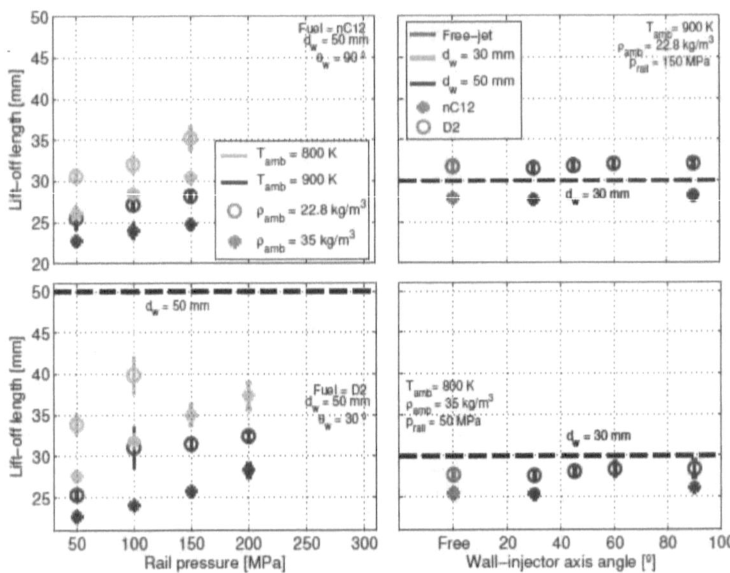

Fig. 6.24: Lift-off length obtained for different parametric changes. Left set: LoL vs injection pressure varying ambient temperature and density. Right set: Variation of LoL vs. wall angle at different wall orientations and fuels (free-jet included). Please note that wall locations are shown in dashed black lines.

visible $LoLs$ at SWI conditions are shown in Figure 6.25 compared to free-jet ones, being approximately unchanged for the two cases. This fact, not only supports that ambient conditions are not changed by the wall, but is consistent enough to consider that LoL does not depend on wall distance (or even its presence) as long as it is located upstream from the wall, and mixing and reaction conditions are met before reaching it. Despite some works describe a shortening in lift-off length with wall distance at high spray momentum conditions [11, 14, 23], they do agree in the use of bowl-like impingement geometries. Nonetheless, the behavior in these researches supports that the main cause of lift-off length reduction in SWI conditions is the recirculation or re-entrainment of hot combustion products that are entrained back into incoming diesel jet, which is not predominant in the present case with a flat wall.

Despite the employed OH* chemiluminescence optical arrangement does not allow to assess what happens for tests points whose free-jet LoL are downstream from the flame thickness location and even less, downstream from the wall, images observed in the frontal natural luminosity setup can help to

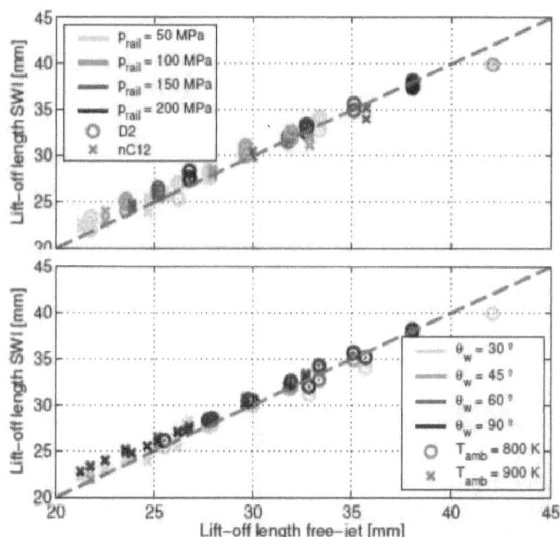

Fig. 6.25: Lift-off length with the quartz wall in SWI conditions vs. ignition delay at free-jet conditions. The gray dashed line represents $ID_{free\text{-}jet} = ID_{SWI}$. Both plots have the same information differently classified. Top: Fuel and injection pressure variation. Bottom: Changes in wall position and gas temperature.

shed light on that matter. Figure 6.26 presents a sequence of frames taken with the frontal camera through the wall for both wall angles 30° and 90°, with the other conditions remaining the same. Tests with those conditions present a free-jet lift-off length of 32.79 mm while in these SWI samples, the wall is set at 30 mm from the injector hole. Differently to the high-temperature and short-LoL point of the sample presented in Figure 4.12, during the steady flame spreading along the wall there is a hole in the 'collision point' where there is no evident chemiluminescence and even the injector tip is visible. This means that SWI not only does not prevent the formation of a lift-off length downstream from the wall, but that a lift-off length zone is formed onto the wall and its radial expansion is approximately constant until the end of the injection.

In order to estimate this zone dimensions, an average image has been made from the quasi-steady flame frames. Then, a threshold-based discretization has been applied similarly as done for spray or flame contour detection (an inverse threshold to take dark zones) and the central 'stain' has been numerically fitted to an ellipse. Frontal visualization through inclined walls could present projected aberration as described for flame penetration in

Figure 4.13, therefore, the metric that has been taken as characteristic dimension for this 'on-wall-lift-off' is the horizontal ratio of the ellipse, referred from now on, to as wall lift-off radius or $WLoR_{NL}$. It is important to highlight that this approach aims to make a qualitative link with the standard definition of lift-off length, but it is still not considered to be a robust variable by the author, as both techniques use different light filters and principles. Therefore, they are not observing the same radical light emission. A variation of this optical setup using an intensified camera for visualization through the wall would be a more consistent experimental methodology to estimate lift-off expansion onto the wall.

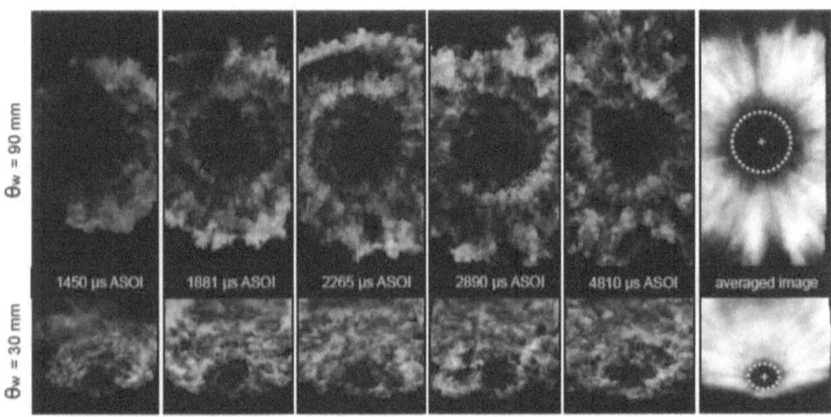

Fig. 6.26: Sequence of the frontal SWI in raw images and definition of wall lift-off ratio $WLoR_{NL}$. ($T_{amb} = 800\,\mathrm{K}$; $\rho_{amb} = 35\,\mathrm{kg\,m^{-3}}$; $p_{rail} = 100\,\mathrm{MPa}$; $d_w = 30\,\mathrm{mm}$; $\theta_w = 90°$; Fuel = D2). The image shows the flame at different times and finally, an average image of SWI phase.

$WLoR_{NL}$ vs. rail pressure is plotted in Figure 6.27 and both wall angle and density are changed. $T_{amb} = 900\,\mathrm{K}$ and nC12 as fuel are high reactivity conditions that do not present free lift-off lengths downstream from the wall, and no free-jet lift-off length of the test matrix reaches the 50 mm wall (the longest free LoL is 48.77 mm). The inclined wall presents a shrunken region with a $WLoR_{NL}$. Although a re-entrainment of hot combustion products has shown to have no effect on visible LoL due to the wall flatness; taking into account the small scale of $WLoR$ and the vicinity of the wall with this reaction zone, it could be considered that wall inclination in the bottom part of the spray acts like a curvature that promotes local hot gases entrainment and therefore, the reduction of the wall lift-off radius. Differently as free-jet *LoL*, changes on wall lift-off radius are stronger with this angle variation than the produced by density and injection pressure. Despite this $WLoR_{NL}$ is not entirely consistent with *LoL*, it is interesting to notice that the trend with rail pressure seems to be linear in both cases. However, a further study measured via OH* chemiluminescence and a test matrix with more large lift-off lengths, would allow not only to know robust values of wall lift-off radius, but to establish a relationship between free *LoL* and *WLoR*. On the other hand, by using the current methodology it is seen that $WLoR_{NL} > LoL_{free} - d_w$, which could suggest that spray-wall interaction is cooling down the reacting spray. This is possible since the wall surface has a lower temperature than the flame (similar to ambient temperature), but the wall is a no-mixing zone with no air entrainment. Nevertheless, the improvement of the technique to establish a link is still needed to support this affirmation with quantitative certainty.

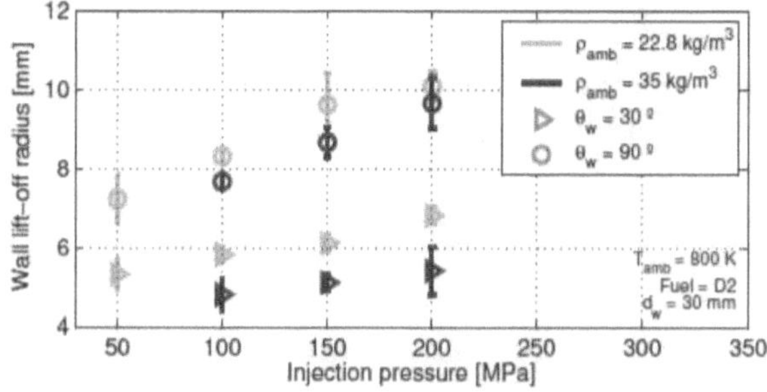

Fig. 6.27: Wall lift-off ratio at different rail pressures, air density and wall angle. (T_{amb} = 800 K; d_w = 30 mm; θ_w = 60°; Fuel = D2).

6.6 Conclusions

A simultaneous visualization with four cameras was made in order to observe reacting sprays morphology in terms of their vapor phase by the use of Schlieren imaging, of the soot-emitted light directly observed by two different views and the lift-off length detected via OH* chemiluminescence. In this campaign, standard air was used inside the chamber so oxygen concentration was approximately 21% for all points. Schlieren images were also used to detect ignition delay with the implementation of an intensity-based processing technique. The use of a quartz wall aimed to allow the flame frontal visualization and additionally, to keep the wall at a temperature similar to the surroundings, and therefore, to reduce the thermal differences on spray-wall and spray-gas interaction.

Ignition delay was seen to be mainly affected by gas temperature and, secondarily by fuel volatility, gas density and rail pressure, similarly as could be expected from well-known behavior of free sprays. Wall angle did not show to have effect on ignition delay. Nevertheless, the different spray distribution after SWI respect to the free-jet situations improves air-fuel mixing and shortens *ID* for normally delayed ignition cases. Consistently, shorter ignition delays are seen with the 30 mm wall distance from the injector. Spray tip penetration and wall spreading are again equivalent to represent spray advance and dependence on test parameters, but are additionally affected by ignition delay in terms of both time of occurrence and premixing time. Then, *R-parameter* is not constant for the whole spray-wall interaction event but it follows five phases: inert spray phase, ignition-induced peak, decelerating valley, momentum rise and a quasi-steady constant value higher than the inert one, behavior observed for free vapor penetration in the literature. Spray thickness behaves similarly to spreading: follows the trends observed in chapter 5 but the effects on ignition delay in spray expansion are additive from the start of combustion.

Sooty flame showed to have the same spreading trends as the spray seen through Schlieren, and additionally it was measured in the horizontal axis thanks to the frontal visualization of the flame through the wall. If flame thickness is measured far from the 'collision point', it is similar regardless of temperature, injection pressure and density; while premixed combustion duration and transitoriness directly affect this thickness in terms of wall geometry parameters. Finally, short lift-off lengths were observed to remain unchanged by the wall. However, OH* chemiluminescence of the very spray thickness covered what happens in wall vicinities. Despite this drawback, frontal natural luminosity images showed that, for long *LoL* conditions, an elliptical hole in the flame is formed onto the wall. This effect was attributed to the radial growth of lift-off length on the plate. This hole was characterized in terms of the horizontal radius and it was described with a new parameter (*WLoR*), which showed to be similarly affected by density and injection pressure as lift-off length, but to be largely reduced by wall inclination as a product of the promoted re-entrainment of burned products into the reaction zone. Nevertheless, these observations are considered not quantitatively robust to establish a link between free *LoL* and *WLoR* due to both metrics were not consistent in terms of optical approach.

References

[1] Bruneaux, Gilles. "Combustion structure of free and wall-impinging diesel jets by simultaneous laser-induced fluorescence of formaldehyde, poly-aromatic hydrocarbons, and hydroxides". In: *International Journal of Engine Research* 9.3 (2008), pp. 249–265 (cited on page 139).

[2] Panão, M.R.O. and Moreira, A.L.N. "Thermo- and fluid dynamics characterization of spray cooling with pulsed sprays". In: *Experimental Thermal and Fluid Science* 30 (2005), pp. 79–96 (cited on page 139).

[3] Mahmud, Rizal et al. "Characteristics of Flat-Wall Impinging Spray Flame and Its Heat Transfer under Small Diesel Engine-Like Condition". In: *SAE Technical Paper 2017-32-0032* (2017) (cited on page 139).

[4] Payri, Raul, Gimeno, Jaime, Peraza, Jesús E., and Bazyn, Tim. "Spray / wall interaction analysis on an ECN single-hole injector at diesel-like conditions through Schlieren visualization". In: *Proc. 28th ILASS-Europe, Valencia* September (2017) (cited on page 139).

[5] Allocca, Luigi, Lazzaro, Maurizio, Meccariello, G., and Montanaro, Alessandro. "Schlieren visualization of a GDI spray impacting on a heated wall: Non-vaporizing and vaporizing evolutions". In: *Energy* 108 (2016), pp. 93–98 (cited on page 139).

[6] Moussou, Julien, Pilla, Guillaume, Rabeau, Fabien, Sotton, Julien, and Bellenoue, Marc. "High-frequency wall heat flux measurement during wall impingement of a diffusion flame". In: *International Journal of Engine Research* (2019) (cited on page 139).

[7] Mayer, Daniel et al. "Experimental Investigation of Flame-Wall-Impingement and Near-Wall Combustion on the Piston Temperature of a Diesel Engine Using Instantaneous Surface Temperature Measurements". In: *SAE International Journal of Engines*. Vol. 7. 3. 2018, pp. 2014-01-1447 (cited on page 139).

[8] Benajes, Jesus, Payri, Raul, Bardi, Michele, and Martí-aldaraví, Pedro. "Experimental characterization of diesel ignition and lift-off length using a single-hole ECN injector". In: *Applied Thermal Engineering* 58.1-2 (2013), pp. 554–563 (cited on pages 141, 161).

[9] Pickett, Lyle M and Siebers, Dennis L. "Soot in diesel fuel jets: effects of ambient temperature, ambient density, and injection pressure". In: *Combustion and Flame* 138.1 (2004), pp. 114–135 (cited on page 141).

[10] Westlye, Fredrik R et al. "Penetration and combustion characterization of cavitating and non-cavitating fuel injectors under diesel engine conditions". In: *SAE Technical Paper 2016-01-0860* (2016), p. 15 (cited on pages 141, 147).

[11] Rusly, Alvin M., Le, Minh K., Kook, Sanghoon, and Hawkes, Evatt R. "The shortening of lift-off length associated with jet-wall and jet-jet interaction in a small-bore optical diesel engine". In: *Fuel* 125 (2014), pp. 1–14 (cited on pages 141, 162).

[12] Payri, Raul, Salvador, Francisco Javier, Gimeno, Jaime, and Peraza, Jesús E. "Experimental study of the injection conditions influence over n-dodecane and diesel sprays with two ECN single-hole nozzles. Part II: Reactive atmosphere". In: *Energy Conversion and Management* 126 (2016), pp. 1157–1167 (cited on pages 144, 160).

[13] Payri, Raul, Garcia-Oliver, Jose Maria, Xuan, Tiemin, and Bardi, Michele. "A study on diesel spray tip penetration and radial expansion under reacting conditions". In: *Applied Thermal Engineering* 90 (2015), pp. 619–629 (cited on pages 145, 147).

[14] Maes, Noud, Hooglugt, Mark, Dam, Nico, Somers, Bart, and Hardy, Gilles. "On the influence of wall distance and geometry for high-pressure n-dodecane spray flames in a constant-volume chamber". In: *International Journal of Engine Research* (2019) (cited on pages 147, 162).

[15] Lipatnikov, A. N., Li, W. Y., Jiang, L. J., and Shy, S. S. "Does Density Ratio Significantly Affect Turbulent Flame Speed?" In: *Flow, Turbulence and Combustion* 98.4 (2017), pp. 1153–1172 (cited on page 152).

[16] Zhang, Ji, Jing, Wei, and Fang, Tiegang. "High speed imaging of OH * chemiluminescence and natural luminosity of low temperature diesel spray combustion". In: *Fuel* 99 (2012), pp. 226–234 (cited on page 154).

[17] Yin, Zenghui, Yao, Chunde, Geng, Peilin, and Hu, Jiangtao. "Visualization of combustion characteristic of diesel in premixed methanol-air mixture atmosphere of different ambient temperature in a constant volume chamber". In: *Fuel* 174 (2016), pp. 242–250 (cited on page 154).

[18] Liu, Haifeng et al. "Time-resolved spray, flame, soot quantitative measurement fueling n-butanol and soybean biodiesel in a constant volume chamber under various ambient temperatures". In: *Fuel* 133 (2014), pp. 317–325 (cited on page 154).

[19] Pang, Kar Mun, Poon, Hiew Mun, Ng, Hoon Kiat, Gan, Suyin, and Schramm, Jesper. "Soot Formation Modeling of n-dodecane and Diesel Sprays under Engine-Like Conditions". In: *SAE Technical Papers* 2015-September.September (2015) (cited on page 160).

[20] Higgins, Brian and Siebers, Dennis L. "Measurement of the Flame Lift-Off Location on DI Diesel Sprays Using OH Chemiluminescence". In: *SAE Paper 2001-01-0918* (2001) (cited on page 161).

[21] Fitzgerald, Russell P, Svensson, Kenth, Martin, Glen, Qi, Yongli, and Koci, Chad. "Early Investigation of Ducted Fuel Injection for Reducing Soot in Mixing-Controlled Diesel Flames". In: *SAE Technical Paper 2018-01-0238* (2018), pp. 1–17 (cited on page 161).

[22] Gimeno, Jaime, Martí-Aldaraví, Pedro, Carreres, Marcos, and Peraza, Jesús E. "Effect of the nozzle holder on injected fuel temperature for experimental test rigs and its influence on diesel sprays". In: *International Journal of Engine Research* 19.3 (2018), pp. 374–389 (cited on page 161).

[23] Pickett, Lyle M and Lopez, Jose Javier. "Jet-wall interaction effects on diesel combustion and soot formation". In: 2005.724 (2005) (cited on page 162).

Chapter 7

Reactive spray against a cooled wall

This last chapter of results covers the observations made from the experimental campaign performed with the steel instrumented wall. In this opportunity, not only diesel combustion is reproduced in the test rig due to ambient gas engine-like thermodynamic conditions, but wall temperature has been controlled to emulate characteristic values that could be found in a piston of an internal combustion engine. Unlike chapter 6, where both wall and ambient temperatures are approximately the same, now $T_{wall} < T_{amb}$ due to wall cooling, which implies that spray-wall heat transfer is simulated and its effects on ignition and spray development can be analyzed. Alongside the optical diagnostics performed in previous chapters for reactive sprays, heat flux was measured by employing high-speed thermocouples fitted in the wall and by the use of a one-dimensional transient wall heat model. Some other 1-D approaches have been followed by other researchers [1–4] and, in the case of the hardware employed in this work, its suitability has been validated as shown in chapter 3.

The test matrix is limited to the use of n-dodecane as only fuel. Instead, two different wall temperatures have been set by regulating the flow of cooling air through the thermo-well system. The wall has been positioned in four different configurations, varying angle and distance respect to the injector tip. Table 7.1 not only shows the controlled conditions for the 108 points of the test plan, but also indicates the location of the thermocouples on the walls,

Table 7.1: Test plan for experiments in the TRI-Wall.

Variable	Values	Units
Wall hardware	None (free-jet) - TRI-Wall	-
Fuel	nC12	-
Injector	Bosch 3-22 ECN Spray D (D103)	-
Energizing time (ET)	2.5	ms
Tip temperature (T_{tip})	363	K
Oxygen fraction (xO_2)	≈0.21	-
Gas temperature (T_{amb})	800 - 900$^{\triangledown}$	K
Gas density (ρ_{amb})	22.8 - 35	kg/m^3
Injection pressure (p_{rail})	50 - 100 - 150	MPa
Wall distance (d_w)	30 - 50	mm
Wall angle (θ_w)	60 - 90	°
Wall surface temperature (T_w)	480$^{\triangledown}$ - 550	K
Total test points	108	points
Thermocouples location$^{\triangle}$ ($\theta_w = 60°$)	TC0: $d_{gc} = 10$ (inline)	mm
	TC1: $d_{gc} = 30$ (inline)	mm
	TC2: $d_{gc} = 18$ (located next to TC0)	mm
Thermocouples location$^{\triangle}$ ($\theta_w = 90°$)	TC0: $d_{gc} = 0$ (centered)	mm
	TC1: $d_{gc} = 15$ (on the side)	mm
	TC2: $d_{gc} = 20$ (inline)	mm

$^{\triangledown}$ These conditions were not tested simultaneously.
$^{\triangle}$ Please refer to Figure 3.10

which change with the wall angle as illustrated in Figure 3.10. It has to be kept in mind that TC0 is a conventional slow thermocouple used as a robust verification of the desired wall temperature, while the probes used for heat flux measured are TC1 and TC2. Three cameras were simultaneously employed in accordance with the setup shown in Figure 4.6, using the configuration that can be found in Table 7.2, similar to the seen in chapter 6.

7.1 Ignition delay

Ignition delay vs. τ_w (both in ASOI reference) for all the points that were measured using the TRI-Wall are shown in two plots in Figure 7.1. The left plot classifies the tests by ambient temperature and density and injection pressure, while right plot shows the same points differenced by wall temperatures, distances and angles. The two clearest trends are obvious from previous chapters: ignition delay is shortened by ambient temperature and the start of spray-wall interaction occurs before if the wall is closer to the

Table 7.2: Details of the optical setup by technique (Figure 4.6).

Feature	Schlieren imaging	N.L. (side)	OH* chemilumin.
Camera	Photron SA-X2	Photron SA5	Andor-iStar
Sensor type	CMOS	CMOS	ICCD
Filter CWL	480 nm	390 nm	310 ± 5 nm
Frame rate	37500 fps	37500 fps	1 frame/inj.
Shutter time	6.49 μs	16.39 μs	-
TTL-Width	-	-	2.5 ms$^{\triangledown}$
Pixels/mm ratio	4.79	4.80	12.11
Reps per point	10	10	10

$^{\triangledown}$The TTL-Delay was set from 1.5 to 3 ms (ASOE) depending on the ignition delay.

injector. Although some other known trends, such as the increasing of τ_w and the shortening of ID with high densities can still be seen, the overall behaviors of the plots could be, in a first glance, considered not to follow clear patterns. In chapter 6, as ignition delay gets longer (takes place close to the wall or after collision), ID_{SWI} deviated from $ID_{free\text{-}jet}$ getting comparatively shorter. However, the resemblance in the thermal interaction between the spray and the surroundings between both free-jet and SWI with the quartz wall, just variating in terms of the effect of the wall on droplets break-up and spray shape evolution, made that most of the general trends remained the same. To shed light on the differences with realistic heat flux between the spray and the cooled wall, the study of ignition delay is divided into three regimes depending

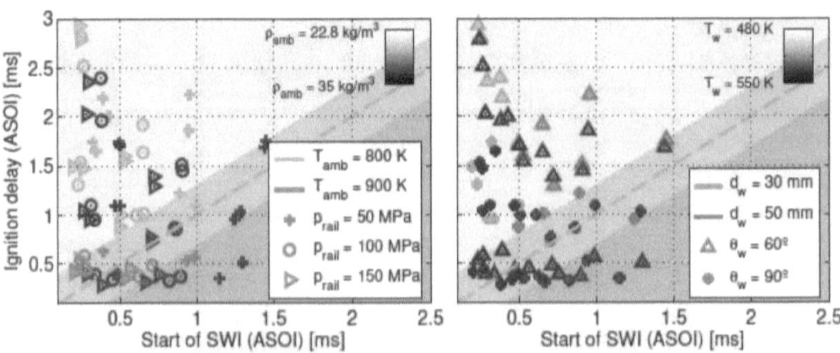

Fig. 7.1: Ignition delay calculated for all SWI conditions vs τ_w and delimitation by ignition location zones (the dashed gray line represents $ID = \tau_w$). Left: Appearance based on operating conditions. Right: Appearance based on wall conditions

on the location of the spray tip when ignition occurs, as shown in the different shades of the plot backgrounds: green for sprays ignited before reaching the wall, light green for ignition given close to the spray-wall impact (between 0.3 ms before and 0.3 ms after τ_w) and yellow for ignition occurring during a well-established spray spreading onto the wall.

Figure 7.2 illustrates ignition delay at different operating conditions and for those different regimes of matching of spray location at ignition time and start of spray-wall interaction. As seen in Figure 7.1, just a few points belong to the 'ignition-before-wall' regime where the spray is still in a free-jet form and the expected trends for that situation remain: ignition delay reduction with higher ambient temperature, densities and rail pressures. If ID and τ_w are similar and the points are in the 'ignition-near-to-wall' region, they seem to have the same trends of ignition delay; nevertheless, the variation with injection pressure seem to be flatter than the one observed in the previous chapter for both free-jet and the quartz wall. Actually, in the situation of 'ignition-on-wall' the trend is changed to have a longer ignition delay with higher injection pressures. This can be explained precisely by the influence of p_{rail} in determining the spray location when it ignites: while injection pressure is known to have a relatively weak effect on ignition delay, it controls the momentum-driven penetration and spreading of the spray. Therefore, once the wall is reached (which happens faster for high injection pressures), the higher the rail pressure, the greater the exposure of the spray to the cold wall. This cools down the still non-reacting spray and delays the ignition occurrence. This effect is also seen with air density which, contrarily to injection pressure, slows down the spray and diminishes spray-wall exposition, and therefore, its trend of shorten ignition delay is intensified in addition to the already known air-fuel mixing improvement. Nonetheless, ambient temperature is still the

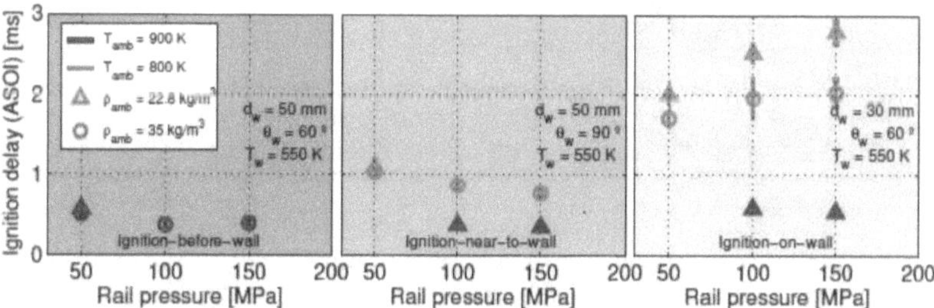

Fig. 7.2: Ignition delay vs. injection pressure for different air thermodynamic conditions and ignition location zones. Left: Tests with ignition before reaching the wall. Center: Tests where ignition takes place close to the wall. Right: Ignition occurs in well defined SWI.

pivotal parameter on defining ignition delay behavior and the effect of other parameters barely exist at high temperature.

On the other hand, the effect of the wall configuration on ignition delay is shown in Figure 7.3, similarly separated by spray tip location at ignition time. By a general observation of the three regimes, it is supported the statement of getting a longer ignition delay as a cause of a more significant exposure to the cooled wall before the ignition event. As it could be expected, in the left plot it can be seen how wall parameters have no effect on ignition delay in free-jet phase. Even when this can be apparently obvious, it is a good validation of how in-chamber conditions remain well-controlled and unaffected by the insertion of the TRI-Wall in the regions that are still far from the thermowell. From 'ignition-before-wall' to 'ignition-on-wall', the strength of parametrical variation effects on ID gradually increases: once SWI is well-established, shorter wall distances imply that the spray enters before in contact with the cold wall, which delays the posterior ignition. Wall temperature effects seem to be negligible. Although a weak trend leads to think that the lower T_w is, the longer the ignition delay, and this behavior could be interpreted as a stronger spray cooling; those differences are too weak to be considered out from the experimental deviation fringe. The author considers that the difference between the two wall temperature conditions of the test matrix (70 K) is still too low to have significant ignition delay gaps, respect to the quartz wall previously studied, with $T_w \approx 800\,\mathrm{K}$ and $T_w \approx 900\,\mathrm{K}$ respectively. This is in accordance with [5], where authors observed slight variations on ignition delay with wall temperatures differing as much as 250 K, and at considerably low ambient densities ($18.03\,\mathrm{kg\,m^{-3}}$), if compared to the test plan of this book.

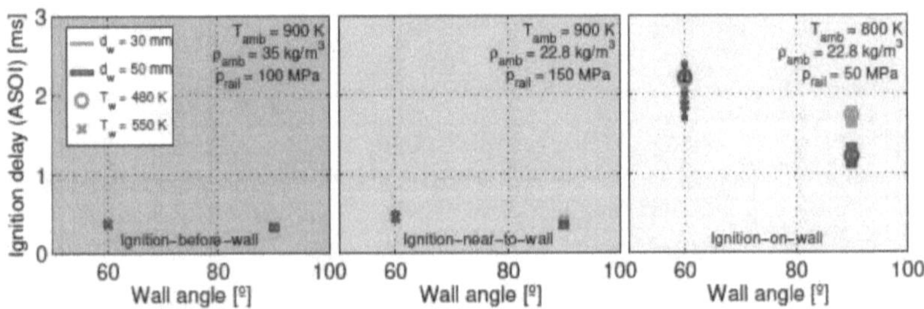

Fig. 7.3: Ignition delay vs. wall angle for different air wall distances, temperatures and ignition location zones. Left: Tests with ignition before reaching the wall. Center: Tests where ignition takes place close to the wall. Right: Ignition occurs in well defined SWI.

A non-obvious but very clear trend is that ignition delay is longer for a wall angle of 60° than for 90°. To illustrate a possible explanation of this, Figure 7.4

shows different Schlieren image backgrounds of the four wall configurations (combinations of two wall distances and two wall angles) for 800 K (top) and 900 K (bottom images). To get these images, the background has been averaged by taking all the frames before SOI of all repetitions independently of rail pressure. This average image allows to largely suppress dark structures caused by the in-chamber density inhomogeneities, characteristic of Schlieren imaging. Green dotted lines in the figure represent the wall edge observed with the HPHTV with no hot flow (valves closed, heaters turned off), which was the original condition in which the wall was set and geometrical features of the image were determined for processing (injector tip and wall location, processing masks, etc).

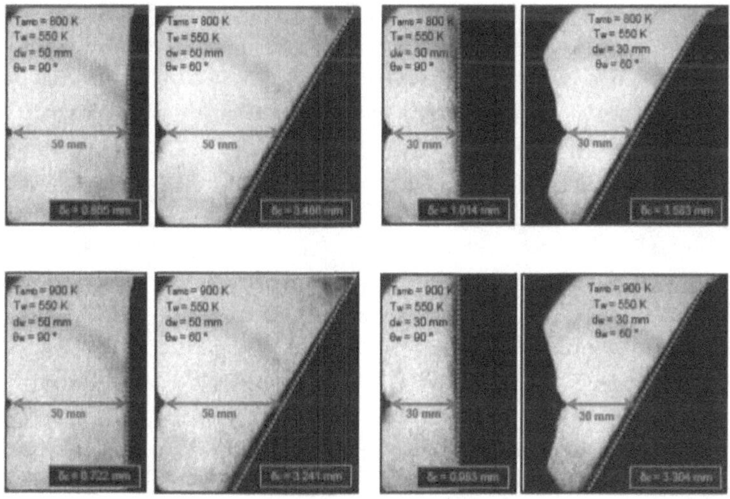

Fig. 7.4: Averaged Schlieren imaging background at $\rho_{amb} = 22.8\,\mathrm{kg\,m^{-3}}$ and various gas temperature and wall conditions. Upper set: $T_{amb} = 800\,\mathrm{K}$. Bottom set: $T_{amb} = 900\,\mathrm{K}$. The green dotted lines represent the original wall edge taken with no hot gaseous flow into the vessel

In the images, a dark layer is seen onto the wall respect to the original position, due to Schlieren light deviation. This layer evidences a strong density (i.e. temperature) gradient in the vicinity of the wall. Similarly as done with the spray, the contour of this layer was obtained, which allowed to obtain a value of the average boundary layer thickness δ_c respect to the original wall surface. This boundary layer is very thin in the perpendicular wall cases, in accordance with the results shown in Figure 4.8 of the chamber temperature characterization, taking into account that the nearest conventional thermocouple used in that study was located at 2 mm from the wall and that, this evaluation could be made only for the 90° walls. The layer for the inclined wall seems to be thicker in the bottom and narrower in the top of the wall, as well as it is significantly thicker than in the perpendicular wall case. As it is known, hot air is less dense than cold air, which makes that in an ambient with temperature gradients, heavier cold air molecules tend to go towards the wall bottom due to gravity and hot air ones to be pushed upwards. This effect is not visible in the free-zone background due to the in-chamber temperature homogeneity of the vessel and the larger and stochastic effect of the flow over the temperature one. Nevertheless, onto the inclined cold wall, the surrounding air is cooled down and instead of just going downwards, the slope of the wall and the air stagnation on it accumulate those cold air particles onto the wall, in a larger extent on the wall bottom due to their weight. This layer of cold air represents through convection a significant contribution to the heat transfer from the air-fuel mixture, delaying it to reach ignition conditions.

Figure 7.5 compares the ignition delay obtained by the use of the two different wall hardware, QT-Wall and TRI-Wall. Short ignition delays are still similar in both cases. The effect of collision-induced break-up is the same for both walls, which makes the larger difference of the wall distance effect to be the produced by the earlier exposure to the cold wall, which delays ignition. This spray-wall contact that causes longer ignition delays is intensified by high injection pressures, in opposition to its known effect that accelerates ignition

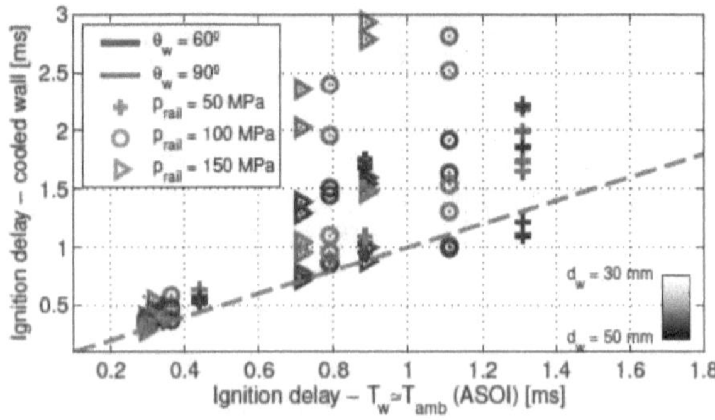

Fig. 7.5: Ignition delay calculated for all tests with the TRI-Wall vs. ignition delay determined in the campaign with the quartz wall.

start. In terms of wall parameters, the strongest parameter that affects *ID* is the wall angle due to the enhance on spray cooling caused by the cold air layer formation in the inclined wall. In comparison to a diesel engine, this nearly quiescent vessel with a static wall may develop a more significant layer. Nevertheless, in automotive engine architectures, it is quite common for spray orientation to be downwards and the bowl geometry could lead to colder air regions.

7.2 Spray evolution on the wall

Figure 7.6 presents a sample of images observed by the use of the Schlieren technique in order to visualize the 'vapor' spray (as mentioned in the previous chapter, gaseous products of combustion are observed and taken inside the contour too). The shown test points are long ignition delay conditions, and

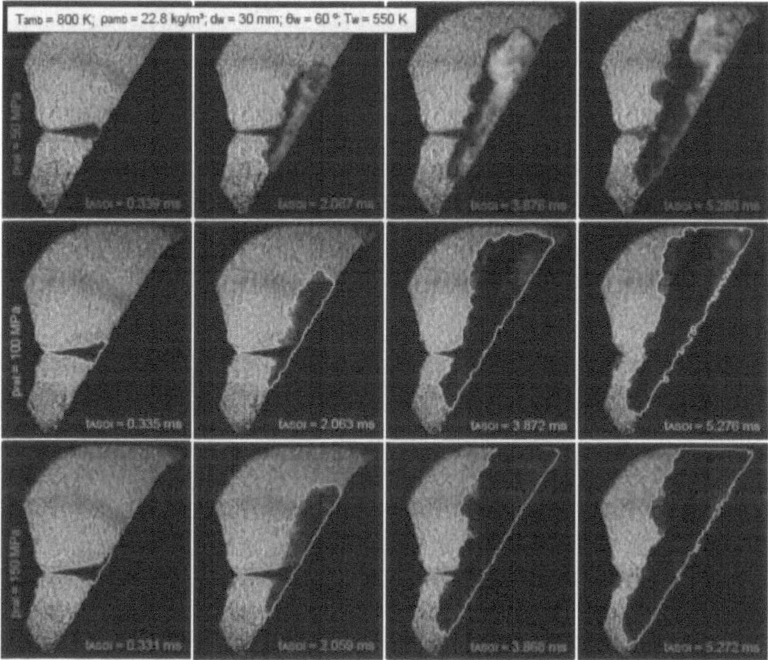

Fig. 7.6: Series of spray images observed via Schlieren varying injection pressure (T_{amb} = 800 K; ρ_{amb} = 22.8 kg m^{-3}; d_w = 30 mm; θ_w = 60°; T_w = 550 K). Top set: p_{rail} = 50 MPa. Center set: p_{rail} = 100 MPa. Bottom set: p_{rail} = 150 MPa.

injection pressure is varied, as seen in the different sets from top to bottom. This particular example is useful to illustrate the phenomenon previously discussed of ignition delay getting longer for higher injection pressures as an effect of the long and prompt exposure of the spray to the cooled wall respect to both flame and ambient temperatures.

In Figure 7.7, spray spreading onto the wall is observed for points at different injection pressures and at T_{amb} = 800 K (left) and 900 K (right), along with their respective *R-parameters*. Which can be seen is a similar behavior to the expected one from the previous chapter: for both before and after ignition, the higher the injection pressure, the higher the spray velocity at the nozzle outlet and the spray momentum flux during both free-jet phase and SWI [6, 7]. Differences with ambient temperature are given by ignition delay in terms of when those variations on spreading happen, and how intense and prolonged they are (more air-fuel premixing time drives to a more vigorous expansion) [8]. It is important to take into account that, in addition to the spray momentum, the crossed effect of injection pressure on ignition delay (spray flow enhanced turbulence against larger spray area in contact with the cold wall) affects spreading. The acceleration phases that characterize reacting diesel spray advancement are still seen here: inert phase, bump when ignition

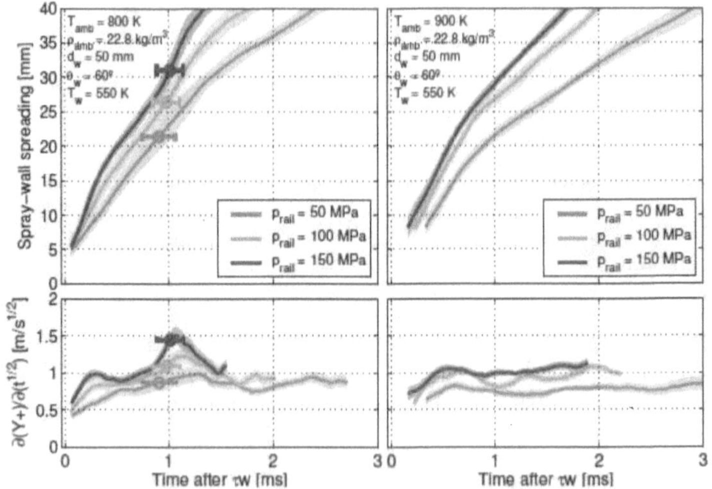

Fig. 7.7: Reacting spray spreading (top) and its respective *R-parameter* (bottom) for different injection pressures and ambient temperatures (ρ_{amb} = 22.8 kg m^{-3}; d_w = 50 mm; θ_w = 60°; T_w = 550 K). Left: Air temperature at 800 K. Right: Vessel set at T_{amb} = 900 K

occurs, deceleration and finally a change in acceleration with a higher rate than in the inert case [9].

On the other hand, Figure 7.8 illustrates wall spreading when both ambient density and wall temperature are changed. First, density promotes gas entrainment into the spray, making its spreading slower, while it shortens *ID* by both representing a higher ambient pressure (at the same gas temperature and oxygen fraction) and by reducing spray exposure to the low temperature wall. This delay on ignition at low ambient densities increments the premixed combustion phase and contributes to the spreading boost respect to the high density environment. Wall temperatures of 480 K and 550 K do not exhibit either significant differences on ignition delay as previously stated, an consequently nor differences on spray spreading. First, having similar ignition delays, there are not variations produced by different pre-mixing levels. Additionally, the thickest mean boundary layers measured are around δ_c = 3.63 mm (being significantly less for perpendicular cases and linearly decreased as go further along the wall). Taking into account that spray thicknesses are in the order of $Z_{th10} \approx 15$ mm and increase at further measuring distances, it can be understood how the density gradient of the thin layer has no important effects on spray advancement onto the wall [10].

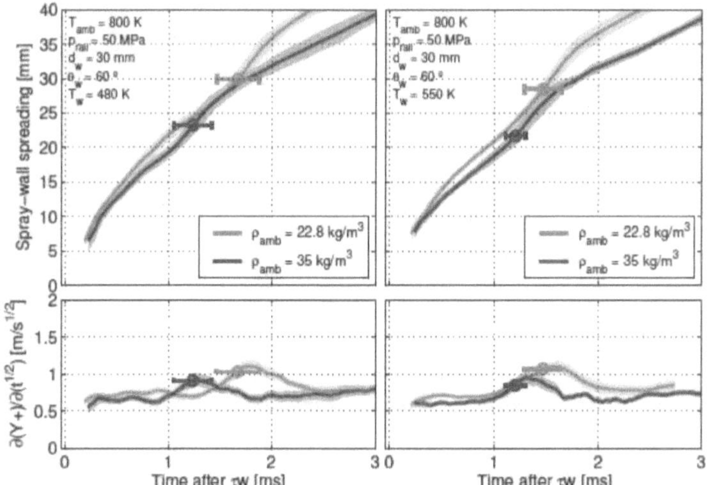

Fig. 7.8: Reactive free penetrations (top) and their *R-parameter* (bottom) for various fuels and air densities (T_{amb} = 800 K; p_{rail} = 50 mm; d_w = 30 mm; θ_w = 60°). Left: Tests at T_w = 480 K. Right: Wall set at 550 K

Different wall distances and inclinations, and their effect on spray spreading is shown in Figure 7.9. Points with short ignition delays are shown in an attempt to reduce the effect of ignition timing. Sprays on the 60° wall are faster in the measured spreading direction, as it is expected. As discussed, the effect of the boundary layer increment with wall inclination (where gases density is locally higher) on spray spreading is negligible. At the same time, the only effect on spreading of wall distance that can be seen in the plots, is given by ignition delay in terms of the changes on τ_w, which is the time reference employed for the horizontal axis, and the crossed effects of a shorter wall distance: a sooner spray opening which provokes a wider exposure to hot and dense air, versus the sooner and longer contact of the spray and the cold wall.

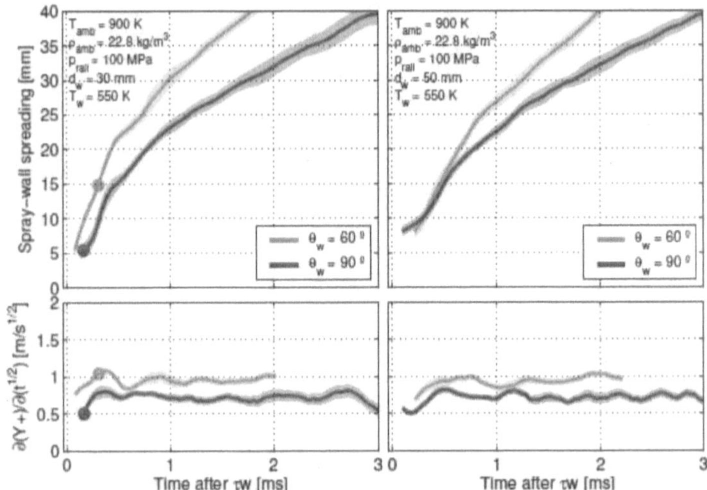

Fig. 7.9: Reactive spreading (top) and their *R-parameter* (bottom) for different wall positions (T_{amb} = 900 K; ρ_{amb} = 22.8 kg m^{-3}; p_{rail} = 100 MPa; T_w = 550 K). Left: Tests with injector-wall distance of 30 mm. Right: Points with the wall located at 50 mm from the injector tip.

Figures 7.10; 7.11 and 7.12 depict the behavior of the vapor thickness of the spray onto the wall at reacting conditions with wall cooling. Figure 7.10 shows in the left the effect of gas temperature and in the right the influence of injection pressure variations. Ambient temperature do not have effect until ignition, when the spray have a general expansion and an increment in its growth rate. As seen, this happens before in the high temperature case but the 800 K point has a longer premixing time. This effect is also seen with injection pressure, which has the aforementioned effects on ignition delay depending on its spreading on the wall and no more representative influences on spray thickness. In the shown case, the most different point in regards to spreading velocity and *ID* (timing and premixing time) is the set at 50 MPa injection pressure, which is directly related to the shown spray thickness. Ignition delay is again the pivotal parameter to define how combustion-driven expansion is provoked, which therefore makes the ambient temperature effect the strongest of all [11, 12].

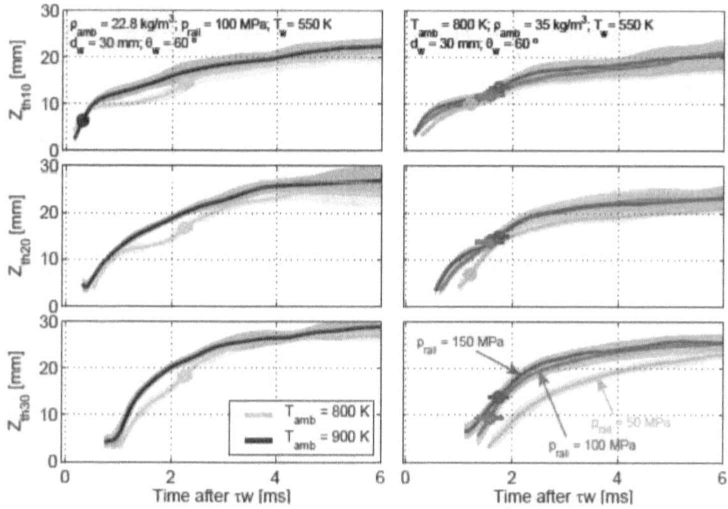

Fig. 7.10: Spray thickness for different ambient temperatures and injection pressures (d_w = 30 mm; θ_w = 60°; T_w = 550 K). Left: Temperature variation at ρ_{amb} = 22.8 kg m^{-3} and 100 MPa. Right: Different injection pressures at ρ_{amb} = 35 kg m^{-3} and 800 K.

The effect of varying both gas density and wall temperature is reflected in Figure 7.11. When gas density is raised, atomization is improved and the spray spreading on the cold wall gets slower, all facts that promote shortening on ignition delay, after that event takes place, still a more intense gas entrainment is given at higher gas density, with a noticeable impact on spray thickness. On the other side, wall temperature does not have a substantial effect on spray thickness as it could be expected since it does not affect ignition delay or spreading, and also by taking into consideration that the effect of gas temperature is only given by the changes on *ID* and not by other mechanisms of SWI.

Different wall positions are illustrated in Figure 7.12, changing injector-wall distance (left) and wall inclination angle (right). In the set of

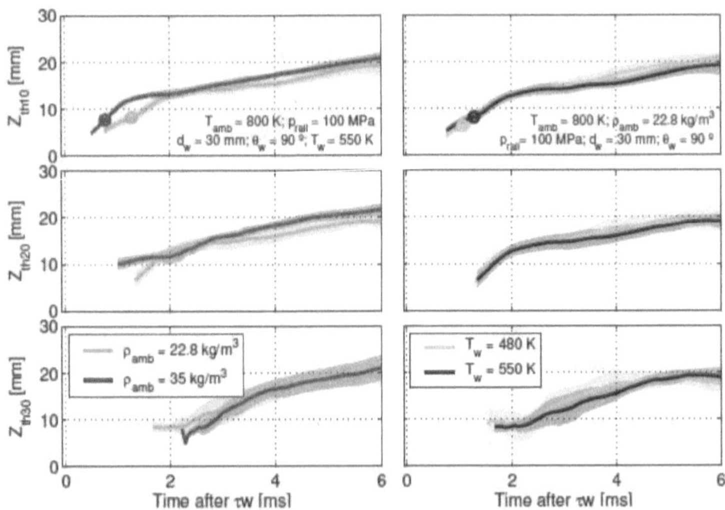

Fig. 7.11: Spray thickness for different gas densities and wall temperatures. ($T_{amb} = 800\,\mathrm{K}$; $p_{rail} = 100\,\mathrm{MPa}$; $d_w = 30\,\mathrm{mm}$; $\theta_w = 90°$). Left: Different gas densities at $T_w = 550\,\mathrm{K}$. Right: $\rho_{amb} = 22.8\,\mathrm{kg\,m^{-3}}$ at two different wall temperatures.

the left, it can be seen how the case is very similar as the observed for gas temperature: there is no substantial effect in the very beginning of the inert phase of thickness. Nevertheless, the ignition delay in the plot is longer for the 30 mm case, due to both the gap in the temporal references of the curves and the sooner spray cooling. However, as seen in previous chapters, thickness is similarly affected by wall distance in regards to both cases having stages with different air entrainment rates. Additionally, in SWIs at non-reacting conditions or $T_w \approx T_{amb}$, the main effect of impingement angle on thickness is given by the transitoriness of the spray-wall interaction due to the deviation of the spray advancement direction. In those cases, this is not well-differenced for various angles since *ID* was unaffected by them. In the present case, the perpendicular wall has the most abrupt change of direction in the spray momentum but, on the other side, ignition gives place to a new transitoriness that is delayed and more intense for the inclined wall with a longer premixing phase.

7.3 Flame morphology visualization

As done in the experiments of the previous chapter, the flame was directly observed via natural luminosity with the same optical setup (please refer to

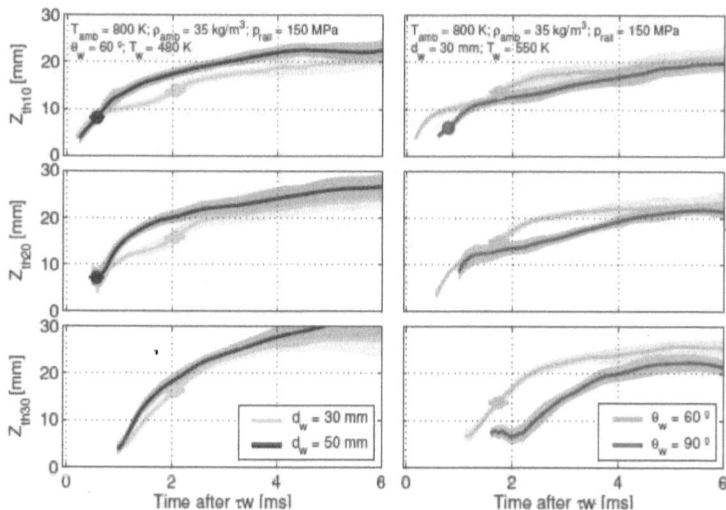

Fig. 7.12: Spray thickness for different wall positions. (T_{amb} = 800 K; ρ_{amb} = 35 kg m^{-3}; p_{rail} = 150 MPa). Left: Different wall distances from the injector tip for θ_w = 90° and T_w = 480 K. Right: Different wall angles fixing d_w = 50 mm and at T_w = 550 K.

Figure 4.6) only from the side, due to the impossibility of the thermo-regulated wall to allow optical accessibility from the front of the HPHTV. However, this side view is useful to determine the evolution with time of the flame behavior and to compare it with the quartz wall experience. A sample of the images that were observed in this campaign is presented in Figure 7.13, where the progression of the flame advance in the chamber is seen, observing how in the case of the figure, the flame luminosity starts being detected near to lift-off length and then, it is extended along the spray until extinction.

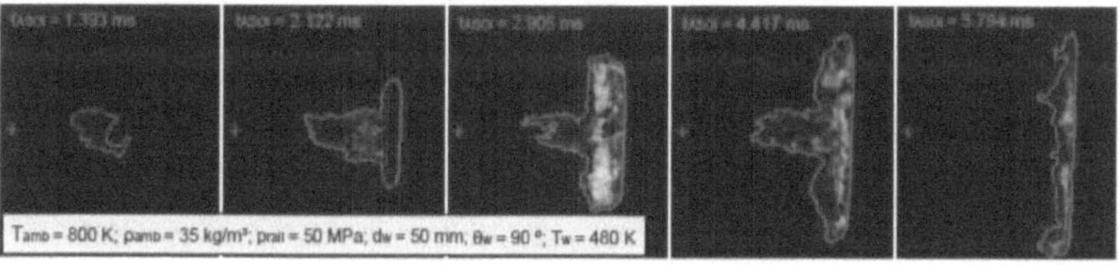

Fig. 7.13: Sample of the flame observed through natural luminosity (T_{amb} = 800 K; ρ_{amb} = 35 kg m^{-3}; p_{rail} = 50 MPa; d_w = 50 mm; θ_w = 90°; T_w = 480 K).

A first analysis is made by observing the effect of ambient temperature and injection pressure on the thickness of the sooty flame in Figure 7.14. As it has been done throughout the manuscript, the shown thickness is measured at 10 mm; 20 mm and 30 mm from the 'collision point'. At first sight, what is noticed is that the thickness profiles seem to be narrower and the vortexes less intense than the observed in chapter 6, finding which will be discussed later with a direct comparison. Figure 7.14 continues showing a thicker flame for high ambient temperature cases, as expected from the enhanced ignition respect to the $T_{amb} = 800\,\mathrm{K}$ point, as well as the longer flame life of the first one due to its sooner start of combustion. In this case at $\theta_w = 60°$, the ignition is so delayed due to the cold air layer formed, that the vortex in the curves of the low temperature point can barely be seen. Nevertheless, the profile is still the same: the tip vortex represented with a peak and then, a semi-stable thickness. Injection pressure, on the other side, still reduces the vortex duration and the general soot thickness observed via natural luminosity, accordingly to [13]. This effect is seen to be intensified as the flame is exposed at higher pressure and velocity to a cold wall, which reduces the overall mixture temperature at the wall vicinities.

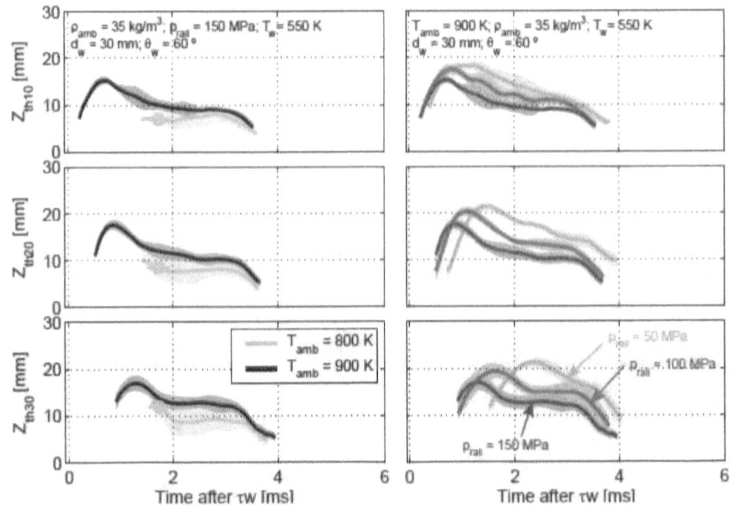

Fig. 7.14: Flame thickness at various ambient temperatures and injection pressures ($\rho_{amb} = 22.8\,\mathrm{kg\,m^{-3}}$; $d_w = 30\,\mathrm{mm}$; $\theta_w = 60°$; $T_w = 550\,\mathrm{K}$). Left: Temperature changed at 150 MPa. Right: Different rail pressures at 900 K.

Figure 7.15 reflects how flame thickness is affected by a change of ambient density and wall temperature. In Figure 7.15-left it is seen how as the spray moves farther onto the wall, not only flame thickness is slightly growing in general, but the frontal vortex is getting a more defined shape in its way. This is shared with the QT-Wall experience, along with the fact of thickness being narrower at lower air densities due to the relative lower gas entrainment rate. The right set of the figure not only shows the two different target T_w of the TRI-Wall, but also the its equivalent point with same parameters and fuel but measured using the quartz wall ($T_w \approx T_{amb}$). Here, the green curve which corresponds to that latter case ($T_w \approx 800\,\mathrm{K}$) and it does not only starts considerably before due to its quicker ignition occurrence, but it also has the previously seen larger and well-defined peak of the flame thickness profile. This Z_{th} is, in all measuring points, decreasing when wall temperature drops. Even while several previous parameters do not seem to be notably affected by wall temperature in the range of control, the flame is narrow enough and in direct contact with the wall to be appreciably affected within the test plan targets (with around a 15% of thickness reduction) in a way that, as seen,

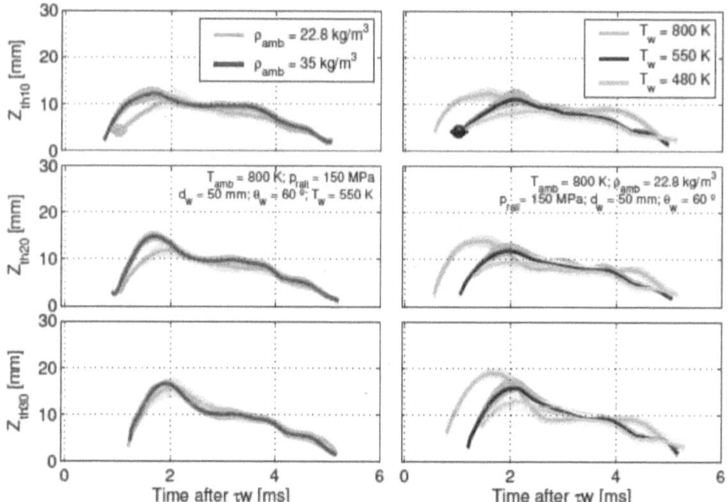

Fig. 7.15: Flame thickness for various gas densities and wall temperatures. ($T_{amb} = 800\,\mathrm{K}$; $p_{rail} = 150\,\mathrm{MPa}$; $d_w = 50\,\mathrm{mm}$; $\theta_w = 60°$). Left: Different gas densities at $T_w = 550\,\mathrm{K}$. Right: $\rho_{amb} = 22.8\,\mathrm{kg\,m^{-3}}$ at three different wall temperatures (including the QT-Wall case with $T_w \approx 800\,\mathrm{K}$).

can be extrapolated and related to the case with wall temperature similar to ambient one.

Different wall distances and angles were set in the test points shown in Figure 7.16. At a constant wall angle and temperature, the effect of placing the plate further from the injector tip makes the flame thicker, in accordance with the results without wall cooling. Nevetheless, in regards to wall angle, the trend seen in chapter 6 of having a progressively narrower flame as the wall is more perpendicular, is not clear in the right set of plots of Figure 7.16, where the differences between points of different wall inclinations is not obvious. In counter-position to the inclination which promotes a thicker flame formation due to its lesser transitoriness and higher spreading velocity, the notably thicker boundary layer, defined as the cold air zone that remains accumulated on the wall, has a similar effect than the aforementioned for wall temperature. In this case, the boundary layer effects are stronger in the inclined wall, respect to the perpendicular wall, where its thickness is about 3 times narrower. It has to be considered that boundary layers of 60° wall cases are commonly as thick as the 34% of the steady (upstream from the front vortex) flame thickness onto the wall. Additionally, in perpendicular wall cases, due to the lateral

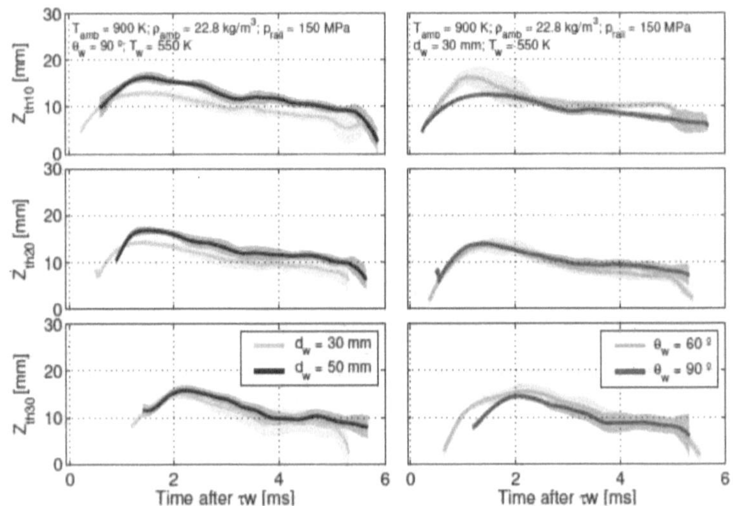

Fig. 7.16: Flame thickness for different wall positions. ($T_{amb} = 900\,\mathrm{K}$; $\rho_{amb} = 22.8\,\mathrm{kg\,m^{-3}}$; $p_{rail} = 150\,\mathrm{MPa}$). Left: Different wall distances from the injector tip for $\theta_w = 90°$. Right: Different wall angles fixing $d_w = 30\,\mathrm{mm}$.

view, the steady thickness after vortex is more covered by the front that is produced by the lateral spreading of the flame.

7.4 Lift-off length

The overall behavior of lift-off length in this campaign is shown in Figure 7.17, where the left set depicts variations on operating conditions such as ambient density, temperature and rail pressure for two different wall settings (top and bottom), while the right set illustrates parametrical variations related to wall characteristics such as its inclination angle in the horizontal axis, wall distance from the injector tip and wall temperature. For the latter, results from the previous chapter where $T_w \approx T_{amb}$ are included, reason to show points with different ambient temperatures in the top and bottom graphs. The strong influence of the thermodynamic conditions of the surrounding air is maintained, shortening *LoL* with density and mainly with temperature increases. It is also shown in the figure how the lift-off length is larger for high injection pressures due to the higher outlet velocities of the fuel.

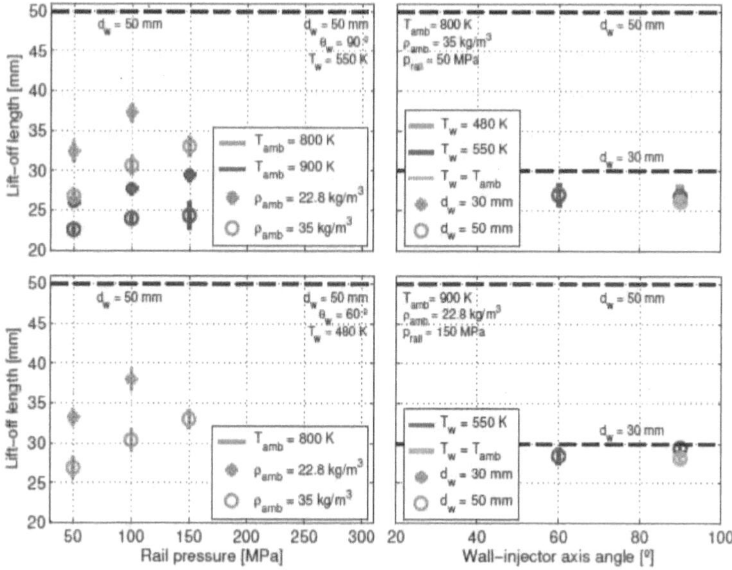

Fig. 7.17: Lift-off length obtained for different parametric changes. Left set: *LoL* vs injection pressure at different gas temperatures and densities. Right set: Variation of *LoL* vs. wall angle at different wall orientations and temperatures (quartz wall included). Please note that wall locations are shown in dashed black lines.

In the previous chapter, it was observed how the wall angle of inclination had no effect on lift-off length for those that are far upstream from the wall. The same trend is seen in Figure 7.17-right for the cooled wall points. Negligible influence on *LoL* from the side of wall temperature in the TRI-Wall cases is also observed, which agrees with the behavior of ignition delay previously shown [14, 15] due to the reduced difference between the target points. This null effect of wall distance or impingement angle on points with short ignition delay, as a lack of re-entrainment of burned products in a flat wall has been reported in other works [16, 17]. Again, it has to be taken into account that the shown points are only the ones with visible lift-off length that are short enough to not being covered by the very thickness of the spray part which is in interaction with the wall. Unfortunately, the architecture of the TRI-Wall makes the frontal view of the spray not optically accessible, which do not allow to obtain quantitative data about lift-off lengths close to the wall or that surpass wall location in free-jet situations. Nevertheless, from this behavior and from the conclusions obtained due to the frontal visualization through the quartz wall, it is expected not only the formation of a wall lift-off radius onto the wall, but also that this radius is larger than the observed when wall and ambient temperatures are similar, as a consequence of the air-fuel mixture cooling which makes its needed conditions of both temperature and droplet vaporization to be reached further downstream.

Still, although the fundamental physical-chemical mechanisms that define lift-off length location seem to remain respect to the wall at the same temperature than ambient, Figure 7.18 is helpful to compare both situations in regards to visible *LoL*. Short lift-off lengths are not affected by the wall, which is logic taking into account that they are far from it and the reacting zone is not affected by its considerably lower temperature, considering that the wall has demonstrated to not be intrusive with the ambient temperature homogeneity into the chamber. Nevertheless, as lift-off length becomes larger, it is noticeable how the values obtained by employing the TRI-Wall are in an increasing deviation respect to the quartz wall ones. This indicates that temperature of the air-fuel mixture for the wall at 50 mm starts to be affected by the cooled wall at around 22 mm from the wall and increasingly more as the *LoL* gets closer. Under this perspective, injection pressure is secondarily prone to enlarge lift-off length due to its trend to push this zone nearer to the cooled wall, while the exact opposite happens with gas density. The effect of T_{amb} is so strong that $LoL_{TRI\text{-}Wall} \approx LoL_{QT\text{-}Wall}$ for all the cases at 900 K ambient temperature. In Figure 7.18, this lift-off length difference between the two hardware is confirmed to be produced by spray cooling observing that, similarly as happens for ignition delay, short *LoL*s are not affected by wall angle while the closest to the wall are slightly larger for the inclined wall, presumably due to the effect of the thicker cold air layer that is formed onto the wall. Moreover, this is an evidence of the greater heat transfer potential the two-phase air-fuel mixture respect to the air into the chamber, whose temperature demonstrated (as seen in Figure 4.8) to remain homogeneous and unaffected by the colder wall temperature except for the region approximately at 3 mm from the wall.

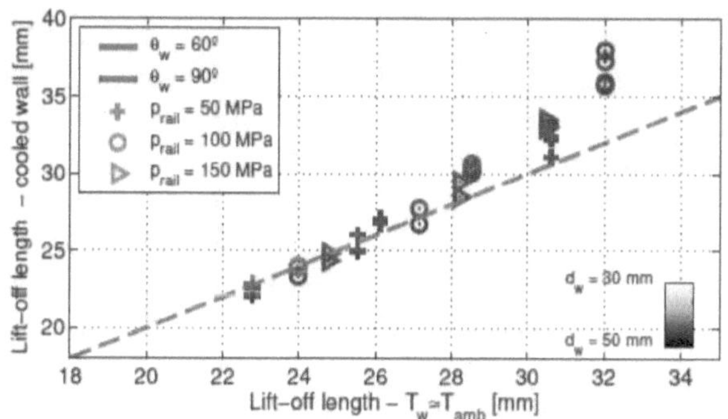

Fig. 7.18: Lift-off length measured for the visible tests with the TRI-Wall vs. quartz wall.

7.5 Flame-wall heat flux

The time resolved wall temperature was registered with the two fast-response thermocouples TC1 and TC2 at two locations per wall. Those locations are different depending on the wall employed, accordingly to the scheme shown in Figure 3.10. The temperature signal is directly used to compute the heat flux transferred to the wall by the flame. Both wall temperature variation ΔT_w respect to the initial one (the set accordingly to the test plan and measured by the thermocouples before SWI) and wall heat flux $\dot{q}_w$ are shown in the following plots for different parametrical variations. In all the following images, the top pair of plots corresponds to the measures made by TC1 and the bottom set comes from TC2.

On the first place, the effect of setting different ambient temperatures is shown in Figure 7.19. As seen, the general profile of wall temperature variation from the test target is a rise that starts with a steep slope when the spray-flame enters in contact with the sensor. This slope gradually decreases to finally reach a peak. After this maximum ΔT_w, wall temperature starts to slowly decrease as a result of flame extinction in that injection event [18,

19]. Although the plot only shows the first 15 ms after the start of spray-wall interaction, it has been verified that the time set between injections (4000 ms) is more than enough to reestablish the wall temperature to its target value. In the right set of Figure 7.19 the profile of the heat flux is observed, where a strong rise is followed by a nearly constant value during the quasi-steady phase of the flame-wall heat transfer, profile that is in agreement with the findings of [1, 20, 21] considering that some of those works took shorter injections in their experiments. Once the flame is extinguished, heat flux dramatically drops to a low level where heat flux is driven by the vapor that remains near the wall in movement, resulting into a higher convective coefficient than the values before SWI. Finally, it can be seen how for the sample that is shown in the figure, the heat flux is higher as the probe is located closer to the geometrical center of the wall defined as 'collision point'. This distance is shown in the upper part of each graph as d_{gc}. The two sensors are reached at post-impact velocities that, as previously discussed, are considerably lower than pre-impingement velocity, and it decreases as the spray advances further from the impingement location.

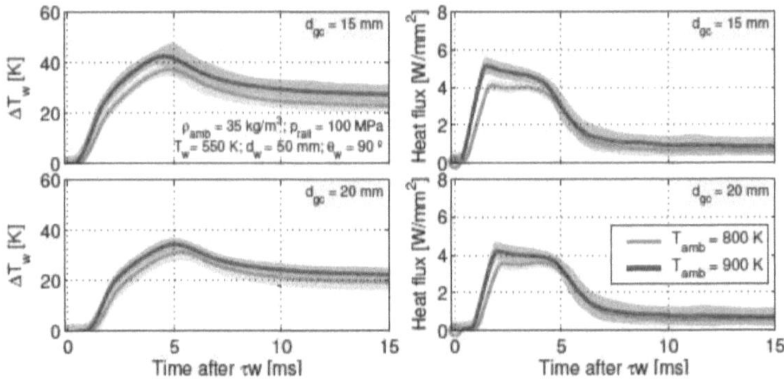

Fig. 7.19: T_w variation (left) and wall heat flux (right) changing T_{amb} (ρ_{amb} = 35 kg m^{-3}; p_{rail} = 100 MPa; d_w = 50 mm; θ_w = 90°; T_{amb} = 550 K).

In regards to the effect of ambient temperature on ΔT_w and $\dot{q}_w$, the higher T_{amb} is, the fuel has in proportion a longer diffusion flame. Higher ambient temperatures lead to higher reactivity and also higher flame temperatures. This represents not only a more significant temperature variation but a stronger heat flux to the wall. Nevertheless, duration from the start of the quasi-steady heat flux to flame extinction is quite similar for both gas temperatures.

Another difference is seen when gas density is changed, as illustrated in Figure 7.20. Ambient density promotes an enhancement on fuel atomization.

It induces not only a higher reactivity but also a stronger gas mass flow in entrainment into the flame during combustion that boosts the energy release to the wall. Wall exposure to a higher air mass flow is also a factor to be taken into account. Furthermore, density affects Reynolds number, which directly influences on the convection coefficient of the hot gases. Figure 7.21 shows the effect of increasing p_{rail}. Turbulence promotes a sooner ignition and the effect on ΔT_w is in a first order, produced by the faster flame spreading onto the wall surface, which affects the convection heat transfer coefficient and finally, increases the wall temperature rise at higher pressures. The same does happen for the measured heat flux, that has a steeper slope and a higher stable value at elevated injection pressures. This quasi-steady phase part of the signal is longer at high injection pressures, due to the relative time when that phase is recorded by the sensor (sooner for high pressure-velocity sprays) and the nearly similar flame extinction.

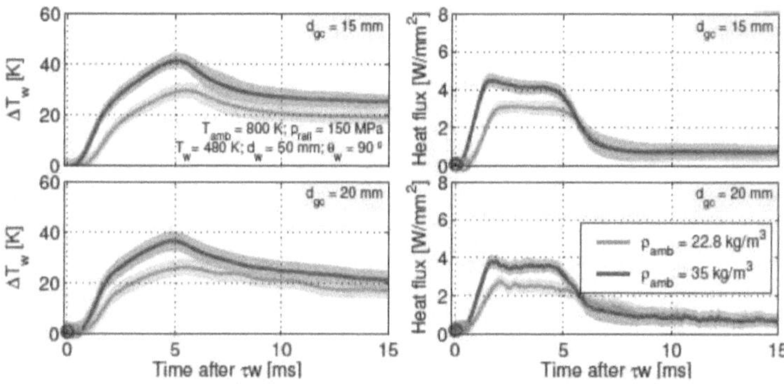

Fig. 7.20: T_w variation (left) and wall heat flux (right) changing ρ_{amb} (T_{amb} = 800 K; p_{rail} = 150 MPa; d_w = 50 mm; θ_w = 90°; T_w = 480 K).

On the other side, in Figure 7.22 it is shown how the target wall temperature has a negligible effect on both its variation and the heat flux, which happens for all points. This, even if could be considered strange, it is in complete accordance with the previously seen in different combustion parameters that would be driven in a large extent by the heat transfer with the wall (such as ignition delay, spray expansion, etc.), except by flame thickness. This can obey different facts: first, as stated before, the difference between the two target wall temperatures T_w is too low to cause a significant variation. It has to be taken into account that, in the case of target gas temperature difference (900 K - 800 K), it is not just larger but it produces a huge increment on flame temperatures [22–24], which is in contact with the

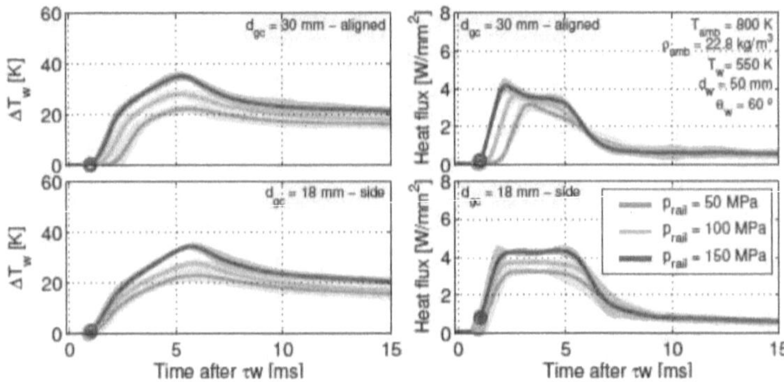

Fig. 7.21: T_w variation (left) and wall heat flux (right) changing p_{rail} (T_{amb} = 800 K; ρ_{amb} = 22.8 kg m^{-3}; d_w = 50 mm; θ_w = 60°; T_w = 550 K).

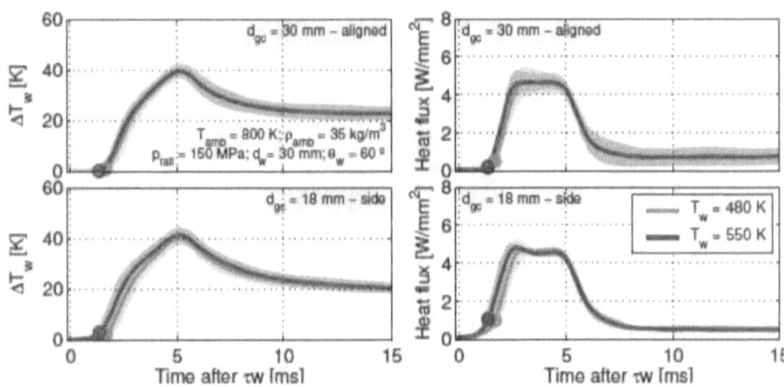

Fig. 7.22: T_w variation (left) and wall heat flux (right) changing T_w (T_{amb} = 800 K; ρ_{amb} = 35 kg m^{-3}; p_{rail} = 150 MPa; d_w = 30 mm; θ_w = 60°).

wall. The difference on those wall temperatures by rising T_{amb} in 100 K is just around 3-5% depending on the point and the thermocouple location. Wall temperature is not the only reduced, but the flame is cooled down too (in a shorter extent), which supports that δT between the flame and the wall is not significantly different for both wall temperatures and, assuming a similar spray velocity from the spreading results, it is expectable to observe similar heat flux by changing target wall temperature.

Figure 7.23 shows variations in surface temperature and heat flux when the wall is at different distances from the injector tip. For a larger d_w, the temperature variation and heat flux seem to be increased for the points shown in the figure and consistently for the whole test matrix. This effect could be out of the expected behavior, taking into account that, as at d_w = 50 mm

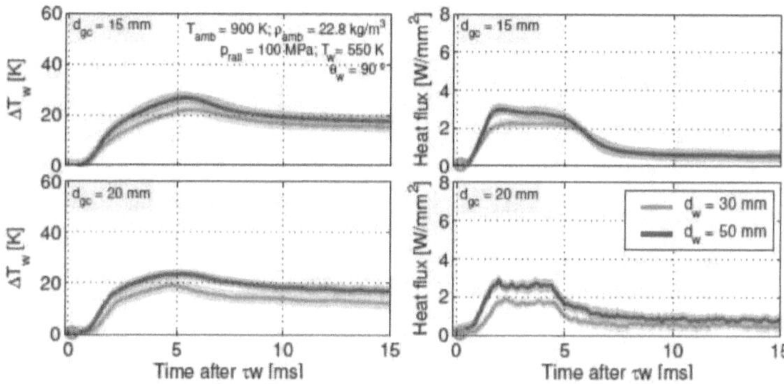

Fig. 7.23: T_w variation (left) and wall heat flux (right) changing d_w (T_{amb} = 900 K; ρ_{amb} = 22.8 kg m^{-3}; p_{rail} = 100 MPa; θ_w = 90°; T_w = 550 K).

the thermocouple is further away from the nozzle, turbulence (i.e. convection coefficient) could be foreseen to be lower. However, from Schlieren results it was observed how since the beginning of SWI, and therefore since the spray tip is located at the 'collision point' or d_{gc} = 0 mm, the spray tip spreading is approximately the same regardless of d_w and then, the convection coefficient may not be largely affected by a significant variation of velocities for sensors located at the same d_{gc} and wall inclination. On the other hand, a possible explanation of the observed behavior of heat flux can come from the temperature distribution inside the flame. It has been observed too how wall distance had negligible effect on visible lift-off length and then, thermocouples are located further downstream from it in the d_w = 50 mm case respect to the 30 mm. The inner gases inside the diffusion flame could be expected to be hotter in locations further from the LoL' region [25–27] and that is something that is also given in the combustion structure of an impinging flame [28, 29]. This factor, together with the low variation on spray velocities at each thermocouple location for different wall distances, could be considered a suitable explanation for the observed trend.

Finally, Figure 7.24 shows both surface temperature rise and heat flux for walls with different inclination angles. It has to be considered that the two fast thermocouples are differently located depending on the wall, as labeled on the top of each graph. That is the first difference between the signals of different wall angles to be taken into account, as it is known that ΔT_w and $\dot{q}_w$ drop in the radial spreading of the spray due to the higher level of spray-sensor contact and the gradual velocity losses. Nevertheless, wall inclination induces a non-homogeneous momentum distribution that favors the upper part of the spray. For the inclined wall, TC1 (top plots) is placed at 15 mm further than for the 90°, but that is compensated in terms of tip velocity by the wall inclination of the first one. Something similar happens for TC2 (bottom graphs) which is similarly distanced from the geometrical center for the two walls (d_{gc} = 18 mm and d_{gc} = 20 mm respectively), what would make the spray velocity higher at the sensor location for the inclined wall. Even when, in the 60° wall, TC2 location is on the side and not aligned in the measured spreading axis, the inclination of spreading in the direction from the center to the thermocouple is still between 60° and 90° (actually 73.4°). These observations on the spreading velocity for each thermocouple are in accordance with the temperature rise observed in Figure 7.24.

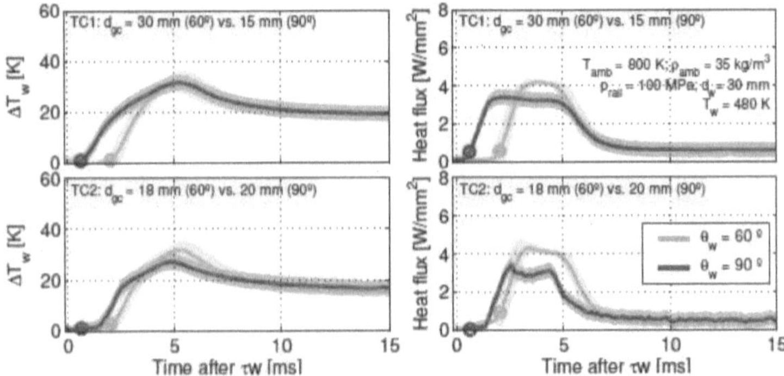

Fig. 7.24: T_w variation (left) and wall heat flux (right) changing θ_w (T_{amb} = 800 K; ρ_{amb} = 35 kg m^{-3}; p_{rail} = 100 MPa; d_w = 30 mm; T_w = 480 K).

In regards to the wall heat flux, the thermal boundary layer strongly delays ignition and, because of that, the duration since the sensor is heated to the decay for flame extinction is shorter for the inclined wall. Nonetheless, this layer can be considered to be displaced from the wall by the high-momentum spray. In the 60° case it is even observed how the spray reaches the sensor pushing hot air at high velocity near to it and then, when ignition takes place, the slope of heat flux is dramatically increase. As expected, the inclined wall reaches higher heat fluxes due to the factor of the spray velocities described above in terms of tip spreading.

7.6 Conclusions

In this chapter, a thermo-controlled and instrumented wall with two fast-response thermocouples has been used to characterize the behavior of spray-wall interaction at engine-like operating conditions. Wall design allowed to test two different wall angles, two distances between the 'collision point' and the injector tip, and two different wall temperatures by regulating the coolant flow in the cold-external face of the wall system. The same operating conditions of previous chapters were tested in this campaign, performing simultaneously flame visualization via natural luminosity, vapor spray visualization recording Schlieren images, lift-off length by the detection of OH* light emission, employing an intensified camera and a 310 nm narrow filter and, finally, time-resolved heat flux calculations from the sensor temperature signals at different points of the wall.

The already known trends for ignition delay were still observed for changes in gas ambient and temperature. For sprays that ignite considerably before reaching the wall, the rest of the behaviors observed are also similar as the expected from the quartz wall study. Nevertheless, the exposure to the cold wall affects the ignition delay variation with parameters: high injection pressures make the spray to spread further onto the wall and increment its contact with it, cooling it down and delaying the ignition. Something similar happens with d_w, which if is reduced implies that the spray is cooled down sooner. A boundary layer due to thermal diffusion was formed onto the wall and it was observed that it gets thicker for the inclined wall, which contributes to cool the spray down and to consequently delay the high temperature chemical reactions. Concerning the geometrical evolution of the spray, its behavior is not different from what was observed from a reacting SWI situation: the spray behaves as an inert spray with the additional influence of ignition delay in terms of timing and intensity of expansion, which is more intense for more premixed combustion conditions.

Flame thickness was measured by visualizing soot natural chemiluminescence. It is observed that, while the other behaviors with parametrical changes do not change respect to the observed with the QT-Wall, there is a notable narrowing of the flame with lower wall temperatures, which also affects the very profile of flame thickness respect to a wall at ambient temperature conditions. Another deviation from the quartz wall experience is observed for different wall angles since, due to the aforementioned boundary layer which becomes thicker for the inclined wall, the 60° wall prevents flame thickness to grow respect to the perpendicular plate.

Lift-off length was laterally observed with an intensified camera. It was found that wall temperature in the range of TRI-Wall target conditions did not affect LoL location. On the other side, compared to the $T_w \approx T_{amb}$ conditions reached with the quartz wall, short lift-off lengths remained unaffected by the hardware while the large ones showed to be even larger for the cooled wall due to the different effective temperature of the air-fuel mixture from determined zone close to the wall.

Finally, transient heat flux transfer by an impinging diesel flame was measured with high temporal resolution. The observed results showed a substantial increment of the heat flux and the wall temperature variation with both ambient temperature and density by increasing flame temperature and gas entrainment. Injection pressure increments the convective coefficient in terms of mixture-flame velocity and turbulence. The effect of spreading velocity is also observed when wall angle is changed.

References

[1] Moussou, Julien, Pilla, Guillaume, Rabeau, Fabien, Sotton, Julien, and Bellenoue, Marc. "High-frequency wall heat flux measurement during wall impingement of a diffusion flame". In: *International Journal of Engine Research* (2019) (cited on pages 171, 191).

[2] Mayer, Daniel et al. "Experimental Investigation of Flame-Wall-Impingement and Near-Wall Combustion on the Piston Temperature of a Diesel Engine Using Instantaneous Surface Temperature Measurements". In: *SAE International Journal of Engines*. Vol. 7. 3. 2018, pp. 2014–01–1447 (cited on page 171).

[3] Meingast, Ulrich, Staudt, Michael, Reichelt, Lars, and Renz, Ulrich. "Analysis of Spray / Wall Interaction Under Diesel Engine Conditions". In: *SAE Technical Paper 2000-01-0272* 724 (2000), pp. 1–15 (cited on page 171).

[4] Panão, Miguel R. O., Durão, Diamantino F. G., and Moreira, António L. N. "Dynamic Spray / Wall Heat Transfer Correlation for 'Cold Combustion' Modes with Port-Fuel Injection". In: *International Review of Mechanical Engineering* (2008) (cited on page 171).

[5] Chen, Beiling et al. "Spray and flame characteristics of wall-impinging diesel fuel spray at different wall temperatures and ambient pressures in a constant volume combustion vessel". In: *Fuel* 235 (2019), pp. 416–425 (cited on page 175).

[6] Payri, Raul, Gimeno, Jaime, Peraza, Jesús E., and Bazyn, Tim. "Spray / wall interaction analysis on an ECN single-hole injector at diesel-like conditions through Schlieren visualization". In: *Proc. 28th ILASS-Europe, Valencia* September (2017) (cited on page 179).

[7] Zhao, Le et al. "Evaluation of Diesel Spray-Wall Interaction and Morphology around Impingement Location". In: 2018 (cited on page 179).

[8] Payri, Raul, Garcia-Oliver, Jose Maria, Xuan, Tiemin, and Bardi, Michele. "A study on diesel spray tip penetration and radial expansion under reacting conditions". In: *Applied Thermal Engineering* 90 (2015), pp. 619–629 (cited on page 179).

[9] Payri, Raul, Salvador, Francisco Javier, Gimeno, Jaime, and Peraza, Jesús E. "Experimental study of the injection conditions influence over n-dodecane and diesel sprays with two ECN single-hole nozzles. Part II: Reactive atmosphere". In: *Energy Conversion and Management* 126 (2016), pp. 1157–1167 (cited on page 180).

[10] Lipatnikov, A. N., Li, W. Y., Jiang, L. J., and Shy, S. S. "Does Density Ratio Significantly Affect Turbulent Flame Speed?" In: *Flow, Turbulence and Combustion* 98.4 (2017), pp. 1153–1172 (cited on page 180).

[11] Gimeno, Jaime, Martí-Aldaraví, Pedro, Carreres, Marcos, and Peraza, Jesús E. "Effect of the nozzle holder on injected fuel temperature for experimental test rigs and its influence on diesel sprays". In: *International Journal of Engine Research* 19.3 (2018), pp. 374–389 (cited on page 182).

[12] Westlye, Fredrik R et al. "Penetration and combustion characterization of cavitating and non-cavitating fuel injectors under diesel engine conditions". In: *SAE Technical Paper 2016-01-0860* (2016), p. 15 (cited on page 182).

[13] Du, Wei, Zhang, Qiankun, Zhang, Zheng, Lou, Juejue, and Bao, Wenhua. "Effects of injection pressure on ignition and combustion characteristics of impinging diesel spray". In: *Applied Energy* 226.February (2018), pp. 1163–1168 (cited on page 185).

[14] Benajes, Jesus, Payri, Raul, Bardi, Michele, and Martí-aldaraví, Pedro. "Experimental characterization of diesel ignition and lift-off length using a single-hole ECN injector". In: *Applied Thermal Engineering* 58.1-2 (2013), pp. 554–563 (cited on page 189).

[15] Higgins, Brian and Siebers, Dennis L. "Measurement of the Flame Lift-Off Location on DI Diesel Sprays Using OH Chemiluminescence". In: *SAE Paper 2001-01-0918* (2001) (cited on page 189).

[16] Maes, Noud, Hooglugt, Mark, Dam, Nico, Somers, Bart, and Hardy, Gilles. "On the influence of wall distance and geometry for high-pressure n-dodecane spray flames in a constant-volume chamber". In: *International Journal of Engine Research* (2019) (cited on page 189).

[17] Rusly, Alvin M., Le, Minh K., Kook, Sanghoon, and Hawkes, Evatt R. "The shortening of lift-off length associated with jet-wall and jet-jet interaction in a small-bore optical diesel engine". In: *Fuel* 125 (2014), pp. 1–14 (cited on page 189).

[18] Pickett, Lyle M and Lopez, Jose Javier. "Jet-wall interaction effects on diesel combustion and soot formation". In: 2005.724 (2005) (cited on page 190).

[19] Magnusson, Alf, Andersson, Sven, and Jedrzejowski, Stanislaw. "Spray-Wall Interaction: Diesel Fuels Impinging on a Tempered Wall". In: *SAE Technical Paper 2006-01-1116* (2006) (cited on page 190).

[20] Zhao, Zhihao, Zhu, Xiucheng, Zhao, Le, Naber, Jeffrey, and Lee, Seong-Young. "Spray-Wall Dynamics of High-Pressure Impinging Combustion". In: 2019 (cited on page 191).

[21] Mahmud, Rizal et al. "Characteristics of Flat-Wall Impinging Spray Flame and Its Heat Transfer under Small Diesel Engine-Like Condition". In: *SAE Technical Paper 2017-32-0032* (2017) (cited on page 191).

[22] Jiotode, Yeshudas and Agarwal, Avinash. "Endoscopic combustion characterization of Jatropha biodiesel in a compression ignition engine". In: *Energy* 119 (2016) (cited on page 192).

[23] Mao, Gongping, Shi, Kaikai, Zhang, Cheng, Chen, Shian, and Wang, Ping. "Experimental research on effects of biodiesel fuel combustion flame temperature on NOX formation based on endoscope high-speed photography". In: *Journal of the Energy Institute* X (2020), pp. 1–12 (cited on page 192).

[24] Jha, Saroj Kumar, Fernando, Sandun, and To, S. D.Filip. "Flame temperature analysis of biodiesel blends and components". In: *Fuel* 87.10-11 (2008), pp. 1982–1988 (cited on page 192).

[25] Flynn, P et al. "Diesel combustion: an integrated view combining laser diagnostics, chemical kinetics, and empirical validation". In: *SAE Paper 1999-01-0509* 724 (1999) (cited on page 194).

[26] Dec, John E. "A Conceptual Model of DI Diesel Combustion Based on Laser-Sheet Imaging". In: *SAE Technical Paper 970873* (1997) (cited on page 194).

[27] Turns, Stephen R. *An introduction to combustion: concepts and applications.* 2000, p. 752 (cited on page 194).

[28] Bruneaux, Gilles. "Combustion structure of free and wall-impinging diesel jets by simultaneous laser-induced fluorescence of formaldehyde, poly-aromatic hydrocarbons, and hydroxides". In: *International Journal of Engine Research* 9.3 (2008), pp. 249–265 (cited on page 194).

[29] Ma, Tianyu, Feng, Lei, Wang, Hu, Liu, Haifeng, and Yao, Mingfa. "Analysis of near wall combustion and pollutant migration after spray impingement". In: *International Journal of Heat and Mass Transfer* 141 (2019), pp. 569–579 (cited on page 194).

Chapter 8

Summary and future works

This chapter is aimed to summarize the main conclusions extracted from the work that has been presented along this book. Moreover, several potential new studies and possible improvements of the present work are listed in order to propose developments that can be made in the future about this topic.

8.1 Summary

This investigation has been conceived considering the prospect of ICEs being present in the foreseeable future, and the consequent need to reduce fuel consumption and harmful pollutants emission to meet the oncoming climate challenges. This book seeks to serve as a contribution to the fundamental understanding of the processes that take place when engine-like injection occurs and the spray interacts with solid surfaces, as a relevant situation for current and future generations of internal combustion engines. Heavy-duty diesel conditions were taken as a particularly interesting topic for this issue. A range of optical and sensor-based diagnostics has been employed in a state-of-the-art facility available at the Universitat Politècnica de València.

A considerable effort h as b een i nvested i nto t he d evelopment and implementation of hardware, experimental techniques and data processing algorithms. This work is well-defined in three experimental campaigns, whose results are reported into their respective chapters based on their specific target: Evaporative and non-reacting spray-wall interaction, reactive SWI considering ignition and combustion using an isothermal wall and finally, spray-wall contact with a wall at engine-like temperature. This progressive evolution from more isolated to more realistic study cases, is considered essential and favorable to understand the underlying mechanisms of spray-wall interaction.

In the non-reacting experiments, parametric variations were carried out for different operating and wall conditions in order to evaluate free and impinged spray behaviors. Free penetration and the spreading along the wall showed to be similarly decreased by ambient density and incremented with injection pressure and to be negligibly affected by gas temperature. In the case of spray spreading, the position of the wall was changed, observing that injector-wall distance did not affect as long as spreading was measured from the start of spray-wall interaction. On the other side, an inclined impingement has a main direction where the spray suffers the least deviation, in which spreading goes faster at more inclination. All those observations were similar for both commercial diesel and n-dodecane, having the last one a faster penetration-spreading due to its lower viscosity. Both steady free penetration and wall spreading, as momentum-driven variables, showed to be proportional to the square root of time, introducing the concept of *R-parameter* as an useful metric to evaluate them, regardless of the temporal reference or any initial condition.

Vapor spray angle was measured, finding not only that density is its strongest influencer but also that it is slightly wider for spray-wall interaction cases respect to free spray. A similar analogy to the penetration-spreading one is the established between spray angle and vapor spray thickness along the wall. Additionally, spray thickness is conditioned to the spray stabilization after impingement and the prior gas entrainment. In regards to the liquid phase of the spray, liquid length showed to be affected mainly by temperature, appreciably by density in terms of atomization improvement, and not to be influenced by injection pressure. Unfortunately, there were few points in the test matrix with effective liquid jet-wall interaction. A liquid spreading could be measured, presenting the same trends as free liquid length and, additionally, being reduced by wall distance and wall angle. That is explained from the similarities on liquid volumes in SWI at the same other conditions. Nevertheless, liquid volume is reduced with the presence of a wall and at shorter wall distances due to the different morphology of the spray which enhances the gas-jet heat transfer.

Two different campaigns were carried out at reacting conditions, both of them with different wall systems. The first one was made with a transparent quartz whose surface was approximately at ambient temperature. Free penetration and spray-wall spreading follow, in a first sight, the same trends observed in the inert atmosphere experiments with injection pressure, ambient density and wall inclination angle. Nonetheless, the spray is expanded due to combustion following the next phases that described in terms of *R-parameter* are: a first zone with non-reacting spray behavior, a peak induced by combustion start, a valley produced by a spray deceleration and finally, a rise in *R-parameter* until a steady value higher than the stable inert one. These produced combustion-induced differences respect to an inert spray can affect both spray penetration and spreading since ignition start, where the ignition delay is pivotal in the reacting spray development: not only this expansion process happens since ignition takes place, but the bump produced in the spray growth profile is more pronounced if ignition is more delayed and therefore, the spray is better mixed before combustion. Higher ambient densities, injection pressures and mainly, gas temperatures shorten ignition delay as expected from literature; wall angle has no effect on ignition delay and short wall distances from the injector tip showed to cause a reduction on long ignition delays, a consistent fact with the effect seen in terms of liquid volume of gas-fuel mixing improvement and with the existence and advancement of spray-wall interaction. Incandescent flame was also directly observed both frontally and from the side. In terms of spreading, the flame follows similar trends and it is quantitatively similar to the observed via Schlieren imaging. However, horizontal spreading is not changed by the wall angle due to the measuring axis projection, which has revealed that spreading dependency on wall angle is just given in the same direction of the inclination.

In regards to lift-off length analysis, short lift-off lengths do not seem to show changes with a wall respect to the free-jet case. OH* chemiluminescence shows to be limited at long *LoLs* due to the probability that lift-off length will be covered by the very spray thickness. However, frontal natural luminosity images are enlightening to observe that the wall does not prevent lift-off length to grow, but it can still be formed onto the wall in form of an elliptical shape, described in this document in terms of its horizontal radius as a new metric (introduced as Wall Lift-off Radius or *WLoR*). This radius, measured from the images of natural luminosity, showed to be susceptible to injection pressure and gas density, as happens for *LoL*, and to wall angle due to the local burned gases re-entrainment into the collision point region. This re-entrainment is negligible out of this zone for a non-confined and flat wall, case in which the spray spreads mainly in the wall direction. The dimensions observed in $WLoR_{NL}$ suggest a reduction in soot levels and the presence of spray cooling respect to the free-jet condition, where wall temperature is still considered nearly the same as ambient. Nevertheless, these observations based on *WLoR* values seek to be careful and not entirely conclusive since optical techniques used in both frontal view and lift-off visualization are not consistent.

The last chapter of results covers the campaign that better emulates engine-like SWI conditions. In this occasion, a stainless steel wall was used and provided with fast-response thermocouples in order to register wall temperature and heat flux from a transient 1-D numerical model, along with the previously mentioned optical diagnostics but frontal natural luminosity. Two target (read before SWI) wall temperatures were set, similar to the ones that can be found in an engine combustion chamber. First, it can be observed how some of the already known *ID* trends from the reacting spray experiments with the quartz wall still remain (i.e. shortening with gas temperature), while its behavior with other parameters like injection pressure, wall distance and wall angle are not globally clear. A criterion was established to define three different regions where the spray is located when ignition takes place, depending on the level of interaction the spray is having with the wall at that instant. From this classification it could be seen that at high injection pressures, the spray spreads more onto the cold wall and it is more exposed to the spray-wall heat exchange, releasing heat and delaying more the ignition start. This trend goes in counterposition to the seen in previous experiences where ignition delay is shortened with injection pressure due to the higher turbulence level. Nevertheless, cooling effect is more significant for well-spread sprays. Similarly, this larger contact between the spray and the wall that delays ignition, is given at shorter wall-injector distances, cases where the spray collides before. Despite there is a significant effect of the lower wall temperature respect to the quartz wall (where $T_w \approx T_{amb}$), the two different controlled wall temperatures for the steel wall did not show major variations on ignition delay. Moreover, a boundary layer of cold air is formed in the wall vicinities. This layer has strongly proven to become thicker for inclined walls due to the gravity-driven accumulation of this cold air in the wall. This phenomenon represents a very pronounced delay on ignition in inclined wall cases respect to the perpendicular ones, due to the defined orientation of the wall into the vessel. These affectations of ignition delay have a direct impact on spray evolution, which is still ruled by the same mechanisms observed in the 'inert' experience and the intensity and timing of the expansion caused by the combustion.

Results that are in accordance with the ignition delay findings obtained with the cooled plate, are the ones from the lift-off length visualization. Visible *LoLs* show to have similar behavior with parametrical changes than the seen in the experiments with the quartz wall: gas density and temperature notably shorten this magnitude while injection pressure locates it further from the nozzle and the wall angle does not affect it, considering that re-entrainment effect is negligible far from the wall. However, the largest lift-off lengths that are still visible with the intensified camera from the side, show to be increased for the cooled wall than for the quartz one, as a consequence of the cooling of the air-fuel mixture in spite of the homogeneity of the chamber ambient temperature, taking more distance from the non-cooled wall as LoL gets closer to the wall. In regards to the flame morphology, it is notably affected by the parameters that promote flame release of heat both via conduction and convection, like wall temperature and wall inclination, in terms of the increase of the boundary layer thickness.

Under diesel-like conditions, the temperature of the designed wall is affected only in the first few millimeters of depth. From 1-D heat flux assumptions, results of heat flux measurements were done. Reactivity of the air-fuel mixture and convective coefficient in terms of spray-flame velocity are pivotal parameters on the heat transfer through the wall. Heat flux and the peak in wall temperature variation are incremented by gas density, temperature and rail pressure. Regarding target wall temperature, the effect on the heat flux observed in the experiments is consistently low, considering as possible causes the predominance of flame temperature at high ambient temperature conditions and the relatively low difference between comparable T_w. As expected, the higher spreading velocity of the flame when the wall is inclined promotes a stronger convection at the thermocouple location and therefore, a higher heat flux than the 90° wall case.

A wide variety of experiments and conditions has been tested, obtaining revealing findings about the behavior of the interaction between diesel sprays with a flat wall. Experiments have been designed to obtain different types of information, gradually going from fundamental understanding at simplified conditions, to a more realistic behavior, where phenomena occurrence can be extrapolated to what happens inside an internal combustion engine. All the data extracted from the efforts invested in this book have been, and still are, used by different researchers from the institutes related to this investigation (academy and industry) to develop, validate and improve their CFD models in environments such as CONVERGE CFD® or StarCCM+®.

8.2 Future directions

The interaction between a fuel spray and a wall in the context of internal combustion engines is quite complex due to the significant transitoriness of the whole process, the large amount of involved mechanisms that strongly depend on the operating conditions and its intrinsically stochastic nature. Despite this book has gathered a large amount of data, this study is by no means enough to achieve a full understanding about the complex phenomena around spray-wall interaction. There are many paths that can be followed to continue the work presented in this manuscript, taking into account the relevance of this topic for powertrain research. Below is listed a series of potential possibilities for future developments that can be performed to assess fruitful conclusions beyond the obtained ones from this work:

- A wider test matrix or, at least, tests based on inert-evaporative spray visualization of points at lower temperatures or shorter wall-nozzle distances, in order to have a large amount of samples with consistent liquid jet-wall interaction. The further study of SWI with the liquid phase of the spray could include the use of the thermo-controlled wall in order to analyze the effect of wall temperature on liquid fuel spreading and the wall-cooling capacity of the spray after exchanging energy with the surrounding, by means of the heat flux measuring procedure. Varying energizing times to characterize injection duration effect on SWI or even taking into account multiple-injection strategies, where each injection may change the initial wall conditions in terms of temperature and deposition formation respect to the previous ones, is an interesting subject of study.

- Droplet-wall and droplet-droplet collision experiments would be a proper addition to the studies with a tempered wall, as a fundamental approach using techniques such as backlit or PDA measurements. On the other hand, the fuel film formed onto the wall after SWI has been found by researchers to drive to UHCs formation, therefore, characterization of post-impingement parameters such as surface wettability, film thickness and distribution at different conditions can be a good addition to this work.

- The employment of a high-speed intensified camera to visualize temporal OH* chemiluminescence would provide not only the evolution of lift-off length, but simultaneous measurements of ignition delay and ignition start location with a single camera and a simpler optical setup. Moreover, experiments with laser-based techniques such as PLIF (Planar Laser-Induced Fluorescence), could be suitable for non-axisymmetrical spray arrangements and it could prevent an optical obstruction of the spray by itself that limit the observation of characteristics such as the thickness profile or long lift-off lengths in SWI situations. Furthermore, soot formation measurements during SWI can be performed in axisymmetrical cases of perpendicular walls or developing techniques to calculate it in inclined impingement conditions.

- The time between injections in the experiments of this work was long enough to have the same boundary and initial conditions repetition-to-repetition and to avoid the deposition of a carbon layer on the wall, preventing both optical contamination of the transparent wall and shot-to-shot variation on the temperature signal registered by the wall fast probes. Nevertheless, at different injection frequencies this layer was observed and, to follow the book scope, prevented. A possible experiment could be to measure the carbon layer thickness and to study its effect on the heat flux through the wall along with other parameters such as ignition delay.

- The heat flux through the inclined cooled wall showed to be affected by the formation of a convection-driven density gradient layer. Since this region seems to be defined by the cold air that remains on the wall due to gravity, this leaves open the study of the influence of the wall orientation (same wall turned upside down or rotated 90° in the thermowell axis) on the formation of that layer, and the spray-wall heat flux.

- Major hardware modifications could give place to more potential experiments. From this book, the following step to realistic experiments is to implement a bowl-shaped wall for both transparent and tempered plates. This supposes a challenge for both optical access through the wall and its cooling respectively. A single-hole injector was the one used in this book. However, jet-jet interaction created on the piston wall after spray spreading is a realistic situation which is really scarce in literature. A radial wall or non-conventional nozzle geometries can be a starting point to conceive this type of campaigns in a similar facility.

- Finally, the capability of heating a plate (the only mean to control wall temperature in this book was to cool down a wall heated by the use of a cold gas) would allow to study situations where $T_{amb} < T_w$ such as the seen in gasoline engines, which are of great interest in current and future powertrain generations. Back to the diesel case, the use of a vessel capable of providing an ambient temperature of 1000 K or 1100 K would require to reconsider the architecture of the thermowell and/or consider the use of liquid coolants such as water or oil, whose convective coefficient was estimated to be too high for the operating conditions of this work.

www.ingramcontent.com/pod-product-compliance
Lightning Source LLC
LaVergne TN
LVHW091412190726
843491LV00006B/1390

* 9 7 8 3 3 8 4 2 3 3 9 2 9 *